T0220681

Measuring ITSM

Measuring, Reporting, and
Modeling the IT Service
Management Metrics that Matter
Most to IT Senior Executives

Randy A. Steinberg

Other books by Randy A. Steinberg:

Implementing ITSM
From Silos To Services - Transforming The IT Organization
To An IT Service Management Valued Partner
Trafford Press ISBN: 978-14907-1958-0

Servicing ITSM
A Handbook Of Service Descriptions For IT Service
Managers and a Means For Building Them
Trafford Press ISBN: 978-1-4907-1956-6

Architecting ITSM
A Reference Of Configuration Items and Building Blocks For
A Comprehensive IT Service Management Infrastructure
Trafford Press ISBN: 978-1-4907-1957-3

Measuring ITSM

Measuring, Reporting and Modeling the IT Service Management Metrics That Matter Most To IT Senior Executives

Order this book online at www.trafford.com
or email orders@trafford.com

Most Trafford titles are also available at major online book retailers.

© Copyright 2013 Randy A. Steinberg.
All rights reserved. No part of this publication may be reproduced,
stored in a retrieval system, or transmitted, in any form or by
any means, electronic, mechanical, photocopying, recording, or
otherwise, without the written prior permission of the author.

Printed in the United States of America.

ISBN: 978-1-4907-1945-0 (sc)
ISBN: 978-1-4907-1944-3 (e)

Cover design by: Anthony Mingo, Design & Art Direction

Because of the dynamic nature of the Internet, any web addresses or
links contained in this book may have changed since publication and
may no longer be valid. The views expressed in this work are solely those
of the author and do not necessarily reflect the views of the publisher,
and the publisher hereby disclaims any responsibility for them.

Any people depicted in stock imagery provided by Thinkstock are models,
and such images are being used for illustrative purposes only.
Certain stock imagery © Thinkstock.

Trafford rev. 11/18/2013

 www.trafford.com

North America & international
toll-free: 1 888 232 4444 (USA & Canada)
fax: 812 355 4082

IT faces a serious challenge . . .

- ✓ It is the only business organization that almost never measures its operational effectiveness and efficiency

- ✓ It seldom measures the costs incurred for the services it delivers outside the budget it is given

- ✓ It monitors technologies but almost never monitors labor in terms of rework, waste and misuse

- ✓ It implements technologies with little measurement of deficiencies and defect rates

In short, IT rarely monitors and manages to the metrics that matter most to IT Senior Executives. Worse yet, the following basic concepts seem to bypass many IT management organizations concerning the services they support and deliver:

"If you don't measure it, you can't manage it"

"If you don't measure it, you can't improve it"

"If you don't measure it, you probably don't care'"

"If you can't influence it, then don't measure it"

IT can no longer continue to operate this way.
It's time to operate IT like a Service Organization.

—The Author

Dedication

This book is dedicated to those very hard working IT professionals, managers and executives who deserve to see their IT solutions deploy and operate day-to-day within acceptable levels of costs and risks to their company.

Table of Contents

Chapter

1

Book Overview

Why This Book Was Written

This book is written in the hopes of helping IT to overcome the metrics gap that exists in many of today's IT organizations. Much has been written lately about how IT needs to be "run like a business". A stronger statement is that IT needs to "act like a business". While technology is certainly important, IT needs to inject much better business management practices by operating services instead of technology silos, measuring the quality and effectiveness of those services and taking timely actions to make sure those services are delivered in line with business needs.

The book focuses on the metrics reporting aspects of running an IT services operation. It assumes that the reader is familiar with ITSM concepts—maybe even in the throes of implementing ITSM practices. Without appropriate metrics, there is no way that those efforts can be validated and their benefits quantified. Without "metrics that matter", it is impossible for any ITSM effort to ever demonstrate its value or operate in a cycle of continuous improvement.

In working with many companies around the globe, it is surprising how many IT professionals really have few ideas on what should actually be measured or how an ITSM metrics program should be run. That is where this book steps in. It attempts to present, in practical terms, how an ITSM Metrics Program can be put together and provides suggested sets of metrics that can be used not only for many ITSM processes but for IT delivered services as well.

There are some very dismal facts about IT and how it is measuring its services. When asked how well business organizations actually measured their IT processes and services, less than 5 percent of respondents are able to say that their ITSM processes are fully measured and even less are measuring their service quality. About half cite "some" measures and less than 25% say they have "absolutely no measures at all". These kinds of results would be totally unacceptable in any other business organization, yet IT has been operating this way for quite some time.

The metrics in this book are suggested recommendations meant as a starting point for any ITSM Metrics Program. It is fully expected that readers will take these and customize them to what fits into their organizations based on unique needs, communication styles and how things are reported. The book attempts to stay as practical as possible. A keen focus is on providing solutions that any organization can implement "starting on Monday morning" versus a lot of metrics theory.

Downloadable Content for This Book

There are a number of tools that you may find helpful related to this book. These can be downloaded at:

http://www.itsmlib.com/downloads/SitePages/Home.aspx

Use the following login:

User ID: **MeasuringITSM@itsmlib.com**
Password: **Downloads2**

The tools are:

Tooling Aid—ITSM Metrics Modeling Tool v3.xls

This tool can be used to measure, report and model the ITSM Metrics that matter. It utilizes many of the concepts discussed throughout this book. It runs as a Microsoft EXCEL® spreadsheet.

Tooling Aid—ITSM Project Modeling Tool v3.xls

This tool can be used to predict the success outcome of any ITSM initiative or improvement effort. It runs as a Microsoft EXCEL® spreadsheet.

Tooling Aid—ITSM Service Metrics Tool.xls

This tool can be used as a dashboard for reporting on the quality of the IT services you are delivering. It runs as a Microsoft EXCEL® spreadsheet.

Book Chapters in Brief

Brief descriptions of remaining book chapters are as follows:

Chapter 2—An Overview of ITSM Metrics

This chapter presents a high level overview of the need for metrics from an ITSM perspective. It covers the difference between typical IT measurements and what metrics matter the most to senior executives.

Chapter 3—The ITSM Metrics Model

This chapter covers basics of a suggested metrics model that can be used with ITSM. It covers basic concepts such as Operational Metrics, Key Performance Indicators (KPIs), Critical Success Factors (CSFs), and other outputs of an ITSM Metrics Program.

Chapters 4 to 17—ITSM Metrics Chapters

These chapters list suggested metrics for each ITSM process. Each chapter covers one ITSM process such as Incident Management, Problem Management, Change Management, and so on. For each ITSM process, a set of operational metrics, KPIs and CSFs is given. Suggested calculations for translating Operational Metrics into KPIs and CSFs are also shown.

Chapter 18—IT Workforce Management Metrics

The measurement of key workforce items such as staff turnover, skill levels, labor waste (non-value labor) cannot be ignored. This chapter covers suggested operational metrics, KPIs and CSFs for this area.

Chapter 19—Measuring Services

This chapter covers suggested metrics for measuring the quality and value of your IT services. It uses balanced scorecard concepts to take a holistic quality view of those services with a focus on customer satisfaction, internal operations, delivery capabilities and financial performance.

Chapter 20—Alternatives If Few Metrics Available

This chapter covers some ideas, tips and approaches that can be used if it is discovered that tools and capabilities are lacking for capturing many IT metrics.

Chapter 21—Using the ITSM Project Modeling Tool

This chapter discusses how a tool can be used to predict the success outcome of your ITSM improvement projects and initiatives.

Chapter 22—Using the ITSM Metrics Modeling Tool

This chapter discusses how a tool can be used to measure, report and model ITSM metrics that matter.

Chapter 23—Implementing an ITSM Metrics Program

This chapter presents an end state picture of how an ITSM Metrics program might operate. It also presents a generalized approach for how such a program might be implemented.

Chapter

2

An Overview of ITSM Metrics

The Importance of ITSM Metrics

IT hates metrics.

IT is about the only business function that that rarely measures itself. Take a look at a manufacturing shop floor. Don't they carefully monitor labor, waste, defects, or efficiency? How about the corporate office? Is there not careful attention to things like stock price, earnings per share, revenue per headcount?

Yet, for some reason, IT has come along these many decades and has little in the way of best practices around measuring what they deliver. This has led to many problems and issues such as:

- Throwing money at ITSM projects with the goal of putting a process in place versus solving a business problem (Maybe thinking problems will go away if only that process was in place—a very risky assumption . . .)

- Inability to justify ITSM initiatives with business management and stakeholders (Management has little respect for initiatives when benefits cannot be seen or measured)
- Poor management decision making (hard to make the right decisions when you are blind to what is going on within your IT operational practices)

A number of IT people may be reading this and thinking something like: "What is this about? We do tons of reporting on metrics. We even issue a major monthly report to management!" This may be so, but the real question here is this:

- Does anyone really read through the report?
- Can management make timely and accurate business decisions based on its content?
- Are the metrics chosen helpful to making decisions or are they just relating historical events that took place?

In short:

Are these the metrics that really matter?

What Are Metrics That Matter?

Having spent much time in the IT organizations of numerous companies, one will find that many IT shops are perfectly comfortable in telling you things like:

- How many IT changes were implemented
- How many incidents of some type occurred
- Current peak utilization of components such as network lines or servers
- How available an application or system was

These are certainly important to know. For now, start to keep them in mind as *Operational Metrics*. However, here are examples of a few painful questions to ask about how well the IT Service Delivery and Support practices really operate:

- How much operational labor is waste versus providing value to the business organization?
- What is your efficiency and effectiveness rate for processing IT changes?
- What is the labor utilization incurred for reacting to incidents and problems?
- What is the defect rate on releases put into production?
- What is the customer impact rate for incidents and problems?

From a management point of view, true *KPIs,* or *Key Performance Indicators,* provide a basis for making business decisions. The previous items are examples of indicators that require a management decision. Some examples using the above:

- Poor efficiency and effectiveness rates may indicate action is needed to reduce wasted labor when changes are being handled and processed
- Additional operational staff may be needed if there will not be enough labor to handle the change workload created by a new application or impending merger
- Incidents and problem rates may be high, but customer impact is low—therefore, IT is doing a fantastic job of protecting services but may not be able to sustain this if business volumes start to increase

In order to get to these decisions, another type of metric is needed to indicate when to take actions. This kind of metric is referred to as a *Tolerance.* This is an indicator that identifies, in advance, the thresholds in which your company expects a KPI to operate and behave.

Organizations may elect to use one or more *Tolerance* metrics to bound KPI results for action decisions. If we use the Customer Impact Rate for Incidents KPI (as an example), an organization may elect to:

- Take no action if the KPI scores an impact rate of less than 10 percent
- Prioritize problem trending activities if the KPI scores an impact rate between 10 and 15 percent
- Establish an availability SWAT team and provide possible customer benefits and incentives if the KPI results score over 15 percent

In the above example, there are two Tolerance metrics in place that correspond to the Customer Impact Rate KPI. These are 10% and 15%. They represent the upper and lower bounds by which actions will be taken if the KPI falls inside or above those bounds.

Hopefully, at this point, you can now begin to see the difference between reporting on a historical operational event (such as "the number of incidents that took place") versus a metric that indicates a decision or action needs to occur.

Resistance to Using IT Metrics

It is not clear how or why IT has evolved over all these years without a solid set of good management measurement practices. Perhaps, historically this may have been the casualty of rapid technology advancement, too much attention on technology for technology's sake or simply too great a divide between IT and the business.

This is something that needs to change now. Extremely poor IT decision making has been taking place, especially with recent IT advances that threatens the very foundations for how IT will ever successfully support the business. Many decisions in terms of off-shoring, cloud computing, reducing staff, investing in new services or projects are being made with a surprisingly low level of facts and information. This is resulting in high levels of failed IT initiatives and solutions that cannot be effectively operated and supported day-to-day. Decisions are highly reactive, being made out of pure management frustration with how IT operates. For senior business executives outside of IT, it boils down to this:

"We have no idea how effectively IT operates and whether IT itself even understands that—the only metric we have to go on is cost and we're taking care of that . . ."

IT simply cannot continue to operate without learning how to effectively govern itself. If this continues, IT service quality will continue to erode at a rapid rate and IT labor and infrastructure operating costs will soar as increasing effort is needed to handle poor IT support and delivery practices.

Here are the common excuses you hear a lot within IT organizations for NOT doing metrics:

Excuse #1:

"We have other priorities . . ."

What priorities? How do you know what they really are? If you're not measuring, how do you know where to prioritize? What makes IT different from every other part of the business that does measure itself? What you are really saying here is *". . . we know what is important for the business and will decide for them without need for any validating facts . . ."* (This is generally not recommended for a healthy career path).

Excuse #2:

"We're very uncomfortable with exposing the levels of our performance with others in the organization . . ."

In other words, there is a desire to keep operating in a manner that IT is comfortable with even though it may be a frustration for everyone in the company outside of IT. Better yet, continue to keep executive management blind to what is going on in IT so they can continue to make poor decisions concerning IT.

Excuse #3:

"We don't have the tools to accurately measure and will wait until management provides funds for this . . ."

In other words, if we can't be perfect, let's not do it at all. Blame management for not providing enough resources. IT is saying: "We can't think of any indicators to measure in the short term and are totally dependent on a tool to tell us what to measure".

Here's an interesting trick you might want to consider:

1. Measure performance based on some small number of indicators that reflect (even if not entirely accurate) key ITSM performance behaviors
2. Issue these in a regular report to senior executives
3. Senior executives will get excited about the kind of results being shown
4. Senior executives will then ask how these could be more accurate or provide further detail
5. IT will tell them further support is needed in the way of a tool
6. Senior executives will fund the tool

Think this doesn't work? Watch what happens the next time a major tool vendor wanders into your organization to sell an IT Dashboard solution. For many executives, that is the first time they will see anything resembling management metrics in IT and they tend to get very excited.

The bottom line here is that without any ITSM measurements that matter, service improvement simply cannot take place. Remember the following guiding principles when encountering any resistance to building and using metrics:

"If you don't measure it, you can't manage it"

"If you don't measure it, you can't improve it"

"If you don't measure it, you probably don't care'"

"If you can't influence it, then don't measure it"

Benefits of Using ITSM Metrics That Matter

There are a many good reasons for building and implementing an ITSM metrics program. Metrics that matter will provide:

- Senior executives and management with indicators from which they can make accurate and timely business decisions
- Visibility into how effectively and efficiently IT support and delivery services truly operate
- A basis for identifying and prioritizing IT service improvement enhancements
- Analytical information to identify service deficiencies and problems before they result in serious impacts
- A process-based focus for getting at root cause of deficiencies in service operations versus finger-pointing and blaming specific workers
- Senior management with confidence that IT is managing itself well

In addition, a set of Metrics That Matter can also be the foundation for modeling the impacts of business and IT decisions. For example, the impact of a business acquisition can be modeled in terms of an increase to specific operational metrics like the number of changes anticipated. This in turn can be calculated into the Change Efficiency Rate to see if it trips past an acceptable Tolerance threshold.

Lastly, the ultimate reason for instituting a Metrics That Matters program is to prevent operational risk. The following outcomes identify the kinds of things that IT should be trying to avoid as part of effective service support and delivery practices:

- ✓ Legal Exposure
- ✓ Service Outages
- ✓ Rework
- ✓ Waste
- ✓ Delayed Solutions
- ✓ Slow Operational Processes
- ✓ Security Breaches
- ✓ Inaccurate Information
- ✓ Slow Turnaround Times
- ✓ Unexpected Costs
- ✓ Higher or escalating costs
- ✓ Low Employee Morale
- ✓ Slow Response to Business Needs and Changes
- ✓ Unwanted PR Exposure
- ✓ Dissatisfied Customers
- ✓ Dissatisfied Suppliers
- ✓ Inability to scale
- ✓ Fines and Penalties
- ✓ High Levels of Non-Value Labor
- ✓ Loss of Market Share
- ✓ Loss of Revenue/Sales

An effective metrics measurement program needs to continually communicate the risk exposure levels to the events shown on the previous page to senior management based on the impacts of business and IT decisions. An approach for doing this is offered by use of the ITSM Metrics Model tool that you can download with this book (See Chapter 1 for Details).

These kinds of communications are critical between IT and the senior leadership of the company. They go a long way towards fostering confidence and ability in IT to manage itself well and proactively take actions based on business needs and priorities.

Chapter

3

The ITSM Metrics Model

Categories of Metrics

The ITSM Metrics Model uses several metric categories that are integrated into an overall metrics framework. These categories are as follows:

- ✓ Operational
- ✓ Key Performance Indicators (KPIs)
- ✓ Tolerances
- ✓ Critical Success Factors (CSFs)
- ✓ Dashboards
- ✓ Outcomes
- ✓ What-Ifs
- ✓ Analytical
- ✓ Other

These categories interact with each other in a manner that translates observations of operational events into indicators that can be used to make key IT and business management decisions. A model of these can be shown as follows:

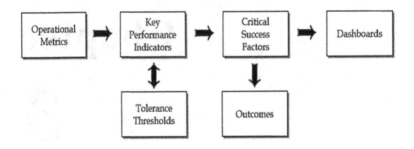

Figure 1: IT Service Management Metrics Model

As shown by the above, *Operational* Metrics will be calculated into *KPIs (Key Performance Indicators)*. *KPI* results will fall into *Tolerance* thresholds. *KPIs* are then calculated into *CSFs (Critical Success Factors)*. *CSFs* are then used to determine *Outcomes (Operational Risks)* and *Dashboard* measures.

Later chapters in this book provide a suggested list of ITSM metrics for each ITSM process for each of these metric categories. The modeling tool that you can download (see Chapter 1 for details) is actually based around this model as well.

The metrics suggested in this book are by no means meant to represent a complete list. They can be used as a starting point for which you may use them as a springboard for identifying additional metrics that your business or IT organization may find useful.

Operational Metrics

These are basic observations of operational events for each ITSM process area. They are the starting point for the model and will be used to calculate the KPIs. Examples of these are as follows:

- Total Number Of Changes Implemented
- Number Of Incidents Reopened
- Number Of Problems In Pipeline
- Number Of Calls Handled
- Customer Satisfaction Ratings
- Total Expended IT Costs

Inputs for these can come from a variety of places such as a Change Management System, Incident Management System, Service Desk ACD Reports, Surveys and other means.

Key Performance Indicators (KPIs)

These are metrics that are used to indicate the performance level of an operation or process. KPIs are used to provide a basis for actionable management decisions. While Operational Metrics are generally historical in nature, KPIs are really the *"Metrics That Matter"*.

KPIs are calculated or derived from one or more Operational Metrics. The results of these calculations are then compared to the Tolerance thresholds to identify whether those results fall within acceptable levels. Examples of KPIs are as follows:

- Change Efficiency Rate
- Change Labor Workforce Utilization
- Incident Repeat Rate
- Capacity Management Process Maturity
- Total Service Penalties Paid

The above examples may not be clearly understood purely by their names. Usually these require small definitions or explanations such that KPIs are fully understood. For this reason, KPIs and their associated calculations should be documented and agreed to by IT and Business Management.

In the above example, the *Change Efficiency Rate* is defined as "*. . . a rate that indicates how many IT Changes in the pipeline for the current month actually were implemented in that month . . .*" In other words, this KPI is meant to identify how efficient the organization is at processing changes. It would be calculated as:

Total Changes Implemented / Total Changes in Pipeline

With the above calculation, the dividend and divisor are both Operational Metrics.

Not all KPIs require a calculation. The Capacity Management Process Maturity, for example, is an Operational Metric (observed from a process audit) and simply carries over as a KPI.

Tolerance Thresholds

Tolerance Thresholds represent upper and lower boundaries for acceptable and non-acceptable KPI values. They should be set by the IT Service Manager and agreed to by IT and Business Senior Management. These are critical, as they form the basis for when management needs to take action or make a key decision.

Each KPI should be associated with one or more Tolerance values. For example, an upper value can represent a desired service target for the KPI and a lower value can represent a warning level or point at which some further action should occur.

The following table shows some examples of KPIs along with their associated Tolerance values:

Table 1: Example KPIs and Their Tolerances

KPI	Service Target	Warning Level
Change Efficiency Rate	92.5%	85.0%
Customer Satisfaction Level	8.7	7.9
Average Incident Resolution Hours	2.0	3.5
Capacity Management Process Maturity	3.0	2.5

In the above example, the service target for the *Change Efficiency Rate* would be 92.5%. Performance of that KPI would be considered acceptable as long as it did not fall lower than 85.0%. If it does fall lower, management actions may need to take place to raise the performance back to acceptable levels.

Note that Tolerance values are based on desired service and performance levels that the business is willing to tolerate. These can vary from one business versus another business.

Critical Success Factors (CSFs)

These are metrics that represent key operational performance requirements which indicate whether a process or operation is performing successfully from a customer or business perspective. They are calculated or derived from one or more KPIs by comparing how those KPIs performed within the tolerance range.

A CSF is usually indicated with a performance level that is indicates a likelihood of success as to whether the CSF was achieved. Typically, this performance level can be something as simple as *High, Medium* or *Low*. Examples of these might look as follows:

Table 2: Example Critical Success Factors

CSF	Performance Level
Protect Services When Making Changes	High
Provide Services At Acceptable Costs	Medium
Continually Improve Availability Of Services	Low

A recommended approach for deriving a CSF is to first identify which KPIs relate to it and then rate the CSF based on the lowest value observed in any one of those KPIs. Using the above for example:

Protect Services When Making Changes might be derived from the Emergency Change Rate, Unauthorized Change Rate and Change Incident Rate KPIs. These all relate to how successful the organization is in meeting that CSF. To receive the performance level of *High*, all KPIs must have met or exceeded their Tolerance threshold targets.

However, let's say one of the KPIs falls below a Tolerance threshold. In this case, the CSF performance level might be

Medium or *Low* depending on how the KPI value fell within the specified Tolerance range for it.

Dashboards

These are key metrics that are represented on a report or graphical interface that indicate the success, at risk or failure of a business operation. They are used to quickly assess the state of operation and take timely actions to correct operational deficiencies. In the ITSM Metrics Model presented here, Dashboard results are derived from CSF results.

Dashboards are generally used by management as a quick and easy way to spot deficiencies without wading through lots of reporting detail. They typically provide information at very high levels and may include drill down capabilities to see things in more detail.

Dashboards come in all forms, shapes and sizes. For the purposes of ITSM "Metrics That Matter", something called the Balanced Scorecard™ is suggested here. The Balanced Scorecard is an approach originally developed in the 1990's by Dr. Robert Kaplan and David Norton from the Harvard Business School. It was originally developed around the concept that financial measures alone are not critical for business success.

The Balanced Scorecard has been around for some time and is generally recognized as an acceptable approach for senior management levels. The scorecard categories recommended for ITSM are:

- ✓ Customer
- ✓ Capabilities
- ✓ Operational
- ✓ Financial
- ✓ Regulatory

Customer

The Customer category represents the customer view of the services being delivered. Are they satisfied? Are they serviced in accordance with agreements and expectations? Examples of some Change Management CSFs that contribute to Customer might be:

- Protect Services When Making Changes
- Make Changes Quickly And Accurately In Line With Business Needs

Both of these CSFs impact how a customer might be receiving (or not receiving) their services.

Capabilities

The Capabilities category represents, in the ITSM sense, the capability of the IT organization to meet business needs. Is there enough capacity to handle planned business volumes? Is there enough capacity to handle anticipated business and IT changes? Does the IT staff possess the right skills? Examples of some Capacity Management CSFs that contribute to Capabilities might be:

- Provide Services With Appropriate Capacity To Match Business Need
- Provide Accurate Capacity Forecasts

The above CSFs represent whether the IT organization is capable of delivering needed capacity to support services by accurately predicting capacity needs and providing needed capacity at the right time to match business requirements.

Operational

The Operational category represents, in the ITSM sense, how well the IT organization is delivering their services on a day-to-day basis. Are service levels being met? Are incidents resolved on a timely basis? Examples of some Incident Management CSFs that contribute to Operational might be:

- Quickly Resolve Incidents
- Maintain IT Service Quality

The above CSFs relate to everyday tasks (in this case Incident Management tasks) and whether those tasks are operating in a repeatable, consistent, efficient and effective manner to quickly resolve incidents and take actions to maintain the quality of the services being delivered.

Financial

The Financial category represents, in the ITSM sense, how well the IT organization is managing and controlling costs as well as protecting and enhancing revenue. Are IT costs effectively managed? Are costs staying within budget? Does revenue received for IT chargeback cover the costs for the services being charged for? Examples of some Financial Management CSFs that contribute to Financial might be:

- Provide Effective Stewardship Of IT Finances
- Maintain Overall Effectiveness Of The IT Financial Management Process
- Recapture IT Costs Through Chargeback For Delivery Of It Services

Regulatory

The Regulatory category represents, in the ITSM sense, how well the IT organization is operating in a manner that protects it against regulatory risks for fines, penalties and audit issues. While not part of the original Balanced Scorecard approach, it has been included here because recent emphasis on IT regulatory issues within the last several years.

Sample questions from this category might include: Is effective stewardship maintained over IT costs? Is the infrastructure protected from unauthorized changes? Is the infrastructure adequately protected from security risks? Examples of some CSFs that contribute to Regulatory might be:

- Provide Effective Stewardship Of IT Finances
- Utilize A Repeatable Process For Handling Changes
- Provide A Repeatable Process For Rolling Out Releases
- Maintain viability of IT Service Continuity Plans

As can be seen from the examples provided, CSFs can contribute to one or more scorecard areas. Likewise, each scorecard area may have one or multiple CSFs.

Outcomes

Outcomes are key indicators of general business risk areas. These are associated with performance indicators that identify the success, at risk or failure of KPIs or CSFs. Outcomes are used to quickly assess the level of risks created by process or operational deficiencies. In short, outcomes are the kind of things that IT is trying to protect against. Examples of these include:

- ✓ Legal Exposure
- ✓ Service Outages
- ✓ Rework
- ✓ Waste
- ✓ Security Breaches
- ✓ Unexpected Costs
- ✓ Slow Response To Business Needs And Changes
- ✓ Unwanted PR Exposure
- ✓ Dissatisfied Customers
- ✓ Fines and Penalties
- ✓ High Levels Of Non-Value Labor
- ✓ Loss of Market Share
- ✓ Loss of Revenue/Sales

Each of the above Outcomes can be associated with a performance indicator such as *High, Medium* or *Low* that might reflect the likelihood of risk that the Outcome will occur. In the ITSM model, the risk level is derived from the mean average of the CSF performance levels.

As an example, the following CSFs relate to risks of incurring *Service Outages*:

- ✓ Quickly Resolve Incidents
- ✓ Minimize The Impact Of Problems (Reduce Incident Frequency/Duration)
- ✓ Protect Services When Making Changes
- ✓ Implement High Quality Releases
- ✓ Protect Services From Capacity Related Incidents

Scoring for an Outcome runs opposite to how the CSFs are calculated. If a CSF scores *Low*, meaning the likelihood of achieving that CSF is low, then the Outcome would score *High*. This means that the risk of the Outcome occurring is high because the CSF achievement was low.

What-Ifs

What-Ifs can be characterized as Use Cases derived from impending business decisions. These will be used to "model" the impacts of those decisions on KPIs and CSFs. An example of a Use Case is simply a scenario for some future event. Examples might include:

- What happens if a major new application goes into production?
- What happens if a planned merger or acquisition occurs?
- What happens if we cut operational staff?

Each of the above is examples of a future event or business decision that IT or business executive management might be thinking of.

For each of these, you can model the impact of such an event by raising or decreasing the values of the Operational Indicators that might be related to them. For example: let's say a new application is going into production. You may decide to model this by:

- ✓ Increasing the Number of Releases in the Pipeline
- ✓ Increasing Labor Hours Spent On Releases
- ✓ Increasing the Number of Changes in the Pipeline
- ✓ Increasing Incidents by some factor such as 30%

Once these changes are made, you can then examine the impact on KPI and CSF results to see if they fall out of Tolerance threshold levels. These kinds of changes reflect your best guess as to how a business decision may impact the quality of your service support and delivery capabilities. Quite a powerful technique to use with executive management!

Analytical

The Analytical category of metrics is used to separate out certain metrics that are really more helpful for supporting research into an issue, incident or service problem. These are metrics that you may report on only on a one-time basis or as part of a drill-down (such as for a Dashboard).

Typically, these kinds of metrics are usually subsets or subdivisions of other metrics. One example might be the Operational Metric of *Total Number of Incidents*. For analytical purposes, you may also wish to see this total broken out by:

- Geographic Region
- Department or Business Unit
- Technology Platform
- IT Service Delivered
- Time of Day, etc.

Recognize that the more of these you include the more complex and labor intensive your ITSM Metrics Program will become. IT frequently makes the mistake of including these in regular reporting to senior management "just in case". This results in a lot of wasted labor in building reports and clouds real management issues that need to be addressed.

Try to avoid as much of this as possible. If pressed to include these, try to keep them in separate reports or as part of a drill-down presentation whenever possible. These are important for when you are looking where to make improvements, but recognize they are important *only* when in the process of making those improvements.

Other

This category is reserved for other kinds of metrics that don't quite fall into the earlier categories. Examples of these might include:

✓ Indicators or other observed events that are being used to represent other metrics that can't be accurately reported on (see Alternatives If Few Metrics chapter for these).

✓ Resource and Service metrics such as server utilizations or transaction response times.

✓ Business metrics such as headcounts, customer and sales volumes or planned revenues

The following chapters in this book contain suggested metrics for many ITSM processes, services, Service Desk and Work Force Management functions. They also show an inventory of recommended Operational, KPI, CSF and associated calculations for each one. These are not meant to be exhaustive, but it is hoped they will kindle ideas within your own organization as to what should be part of your ITSM metrics program.

While not an exhaustive list of metrics by any means, it is strongly felt that if only the metrics listed in this book were tracked and modeled, simple as they are, IT will still be greatly much better off than where many organizations stand today.

Chapter

4

Incident Management Metrics

Operational Metrics

The following table lists suggested Incident Management Operational Metrics.

Table 3: Incident Management Operational Metrics

XREF	METRIC
A	Total Number Of Incidents
B	Average Time To Resolve Severity 1 and Severity 2 Incidents (Hours)
C	Number Of Incidents Resolved Within Agreed Service Levels
D	Number Of High Severity/Major Incidents
E	Number Of Incidents With Customer Impact
F	Number Of Incidents Reopened
G	Total Available Labor Hours To Work On Incidents (Non-Service Desk)
H	Total Labor Hours Spent Resolving Incidents (Non-Service Desk)
I	Incident Management Tooling Support Level
J	Incident Management Process Maturity

Suggested sources for Incident Management Operational Metrics can be found in places such as:

- ✓ Incident Management System Reports
- ✓ Labor or Other HR Reports
- ✓ Process and Tool Assessment Audit Findings

Key Performance Indicators (KPIs)

The following table lists suggested KPIs and how they are calculated from the Operational Metrics listed previously.

Table 4: Incident Management KPIs

XREF	KPI	CALCULATION
1	Number Of Incident Occurrences	A
2	Number Of High Severity/ Major Incidents	D
3	Incident Resolution Rate	C/A
4	Customer Incident Impact Rate	E/A
5	Incident Reopen Rate	F/A
6	Average Time To Resolve Severity 1 and Severity 2 Incidents (Hours)	B
7	Incident Labor Utilization Rate	H/G
8	Incident Management Tooling Support Level	I
9	Incident Management Process Maturity	J

Why These Metrics (KPIs) Matter

The KPIs described earlier are critical to managing and controlling Incident Management activities. The following table lists each Incident Management KPI and the question it is trying to answer:

Table 5: Incident Management KPI Performance Points

KPI	Question Being Answered
Number Of Incident Occurrences	How many incidents did we experience within our infrastructure?
Number Of High Severity/Major Incidents	How many major incidents did we experience?
Incident Resolution Rate	How successful are we at resolving incidents per business requirements?
Customer Incident Impact Rate	How well are we keeping incidents from impacting customers?
Incident Reopen Rate	How successful are we at permanently resolving incidents?
Average Time To Resolve Severity 1 and Severity 2 Incidents (Hours)	How quickly are we resolving incidents?
Incident Labor Utilization Rate	How much available labor capacity was spent handling incidents?
Incident Management Tooling Support Level	How well does our current tool set support Incident Management activities?
Incident Management Process Maturity	How well do we execute our Incident Management practices?

Critical Success Factors (CSFs)

The table below lists suggested Critical Success Factors for Incident Management. The KPI references listed in the right column indicate which KPIs are used as input for the associated CSF.

Table 6: Incident Management CSFs

CSF	KPI
Quickly Resolve Incidents	5,6,8
Maintain IT Service Quality	1,2,3,4,8,9
Improve IT And Business Productivity	7,8
Maintain User Satisfaction	4,8,9

Chapter

5

Problem Management Metrics

Operational Metrics

The following table lists suggested Problem Management Operational Metrics.

Table 7: Problem Management Operational Metrics

XREF	METRIC
A	Number Of Repeat Incidents
B	Number Of Major Problems
C	Total Number Of Incidents
D	Total Number Of Problems In Pipeline
E	Number Of Problems Removed (Error Control)
F	Number Of Known Errors (Root Cause Known and Workaround In Place)
G	Number Of Problems Reopened
H	Number Of Problems With Customer Impact
I	Average Problem Resolution Time—Severity 1 and 2 Problems (Days)
J	Total Available Labor Hours To Work On Problems

XREF	METRIC
K	Total Labor Hours Spent Working On And Coordinating Problems
L	Problem Management Tooling Support Level
M	Problem Management Process Maturity

Suggested sources for Problem Management Operational Metrics can be found in places such as:

- ✓ Incident Management System Reports
- ✓ Problem Management System Reports
- ✓ Labor or Other HR Reports
- ✓ Process and Tool Assessment Audit Findings

Key Performance Indicators (KPIs)

The following table lists suggested KPIs and how they are calculated from the Operational Metrics listed previously.

Table 8: Problem Management KPIs

XREF	KPI	CALCULATION
1	Incident Repeat Rate	A/C
2	Number Of Major Problems	B
3	Problem Resolution Rate	E/D
4	Problem Workaround Rate	F/D
5	Problem Reopen Rate	G/D
6	Customer Impact Rate	H/D
7	Average Problem Resolution Time—Severity 1 and 2 Problems (Days)	I
8	Problem Labor Utilization Rate	K/J
9	Problem Management Tooling Support Level	L
10	Problem Management Process Maturity	M

Why These Metrics (KPIs) Matter

The KPIs described earlier are critical to managing and controlling Problem Management activities. The following table lists each Problem Management KPI and the question it is trying to answer:

Table 9: Problem Management KPI Performance Points

KPI	Question Being Answered
Incident Repeat Rate	How effective are we at minimizing repeat incidents?
Number Of Major Problems	How many major problems did we experience?
Problem Resolution Rate	What percentage of problems have we eliminated?
Problem Workaround Rate	For what percentage of problems did we implement workarounds?
Problem Reopen Rate	How successful are we at removing problems permanently?
Customer Impact Rate	How well are we keeping problems from impacting customers?
Average Problem Resolution Time— Severity 1 and 2 Problems (Days)	How long does it take us to resolve problems?
Problem Labor Utilization Rate	How much available labor capacity was spent handling problems?
Problem Management Tooling Support Level	How well does our current tool set support Problem Management activities?
Problem Management Process Maturity	How well do we execute our Problem Management practices?

Critical Success Factors (CSFs)

The table below lists suggested Critical Success Factors for Problem Management. The KPI references listed in the right column indicate which KPIs are used as input for the associated CSF.

Table 10: Problem Management CSFs

CSF	KPI
Minimize Impact Of Problems (Reduce Incident Frequency/Duration)	1,2,4,6,7
Reduce Unplanned Labor Spent On Incidents	1,3,4,5,8,9
Improve Quality Of Services Being Delivered	2,6
Resolve Problems and Errors Efficiently and Effectively	3,4,5,7,8,9,10

Chapter

6

Request Fulfillment Metrics

Operational Metrics

The following table lists suggested Request Fulfillment Operational Metrics.

Table 11: Request Fulfillment Operational Metrics

XREF	METRIC
A	Total Number Of Requests
B	Number Of Requests Fulfilled Within Agreed Service Levels
C	Number Of Requests Fulfilled Without Service Desk Escalation
D	Number Of Requests Fulfilled Without Human Intervention
E	Number of Requests Fulfilled With Proper Authorization
F	Total Number Of Requests That Have Been Fulfilled
G	Total Number Of Requests That Caused Incidents

XREF	METRIC
H	User Satisfaction Levels For Request Fulfillment
I	Total Available Labor Hours To Work On Requests (Non-Service Desk)
J	Total Labor Hours Spent Fulfilling Requests (Non-Service Desk)
K	Request Fulfillment Tooling Support Level
L	Request Fulfillment Process Maturity

Suggested sources for Request Fulfillment Operational Metrics can be found in places such as:

- ✓ Request Management System
- ✓ Service Catalog System
- ✓ Incident Management System
- ✓ User Satisfaction Ratings
- ✓ Staffing Reports
- ✓ Labor Reports
- ✓ CMMI Tool Ratings or Tool Surveys
- ✓ Process Assessment Results

Key Performance Indicators (KPIs)

The following table lists suggested KPIs and how they are calculated from the Operational Metrics listed previously.

Table 12: Request Fulfillment KPIs

XREF	KPI	CALCULATION
1	Request Fulfillment On-Time Delivery Rate	B/F
2	Request Fulfillment First Call Rate	C/F
3	Request Automation Rate	D/F
4	Request Authorization Rate	E/F
5	Request Processing Rate	F/A
6	Request Incident Rate	G/A
7	Request Fulfillment User Satisfaction Rate	H
8	Request Labor Utilization Rate	J/I
9	Request Fulfillment Tooling Support Level	K
10	Request Fulfillment Process Maturity	L

Why These Metrics (KPIs) Matter

The KPIs described earlier are critical to managing and controlling Request Fulfillment activities. The following table lists each Request Fulfillment KPI and the question it is trying to answer:

Table 13: Request Fulfillment KPI Performance Points

KPI	Question Being Answered
Request Fulfillment On-Time Delivery Rate	How successful are we at fulfilling requests on time?
Request Fulfillment First Call Rate	How many requests can be immediately fulfilled by the service desk?
Request Automation Rate	How successful are we at automating fulfillment of requests?
Request Authorization Rate	How successful are we at fulfilling only requests that are authorized?
Request Processing Rate	How successful are we at processing requests in our pipeline?
Request Incident Rate	Do activities for fulfilling requests cause incidents?
Request Fulfillment User Satisfaction Rate	How satisfied are users with the handling of their requests?
Request Labor Utilization Rate	How much available labor capacity was spent fulfilling requests?
Request Fulfillment Tooling Support Level	How well does our current tool set support Request Fulfillment activities?
Request Fulfillment Process Maturity	How good are our Request Fulfillment practices?

Critical Success Factors (CSFs)

The table below lists suggested Critical Success Factors for Request Fulfillment. The KPI references listed in the right column indicate which KPIs are used as input for the associated CSF.

Table 14: Request Fulfillment CSFs

CSF	KPI
Requests fulfilled in a manner aligned to agreed service level targets	1
Only authorized requests are fulfilled	4
Maintain User Satisfaction	1,6,7
Requests are fulfilled efficiently and effectively	1,2,3,5,6,8,9,10

Chapter

7

Event Management Metrics

Operational Metrics

The following table lists suggested Event Management Operational Metrics.

Table 15: Event Management Operational Metrics

XREF	METRIC
A	Total Number Of Incidents
B	Number Of Incidents Raised By Customer Contacts
C	Number Of Incidents Triggered By Events
D	Total Number Of Events Generated
E	Number of Duplicate Events
F	Number of Events Without Human Intervention
G	Total Number of CIs
H	Number of CIs Not Being Monitored
I	Average Event Response (or Acknowledge) Time (Seconds)
J	Event Management Tooling Support Level
K	Event Management Process Maturity

Suggested sources for Event Management Operational Metrics can be found in places such as:

- ✓ Incident Management System
- ✓ Event Management System
- ✓ Configuration Management System
- ✓ CMMI Tool Ratings or Tool Surveys
- ✓ Process Assessment Results

Key Performance Indicators (KPIs)

The following table lists suggested KPIs and how they are calculated from the Operational Metrics listed previously.

Table 16: Event Management KPIs

XREF	KPI	CALCULATION
1	Incident Customer Impact Ratio	B/A
2	Percent Of Incidents Triggered By An Event	C/A
3	Event Noise Ratio	E/D
4	Event Automation Ratio	F/D
5	Event Monitoring Coverage Ratio	1-(H/G)
6	Average Event Delivery Time (Seconds)	I
7	Event Management Tooling Support Level	J
8	Event Management Process Maturity	K

Why These Metrics (KPIs) Matter

The KPIs described earlier are critical to managing and controlling Event Management activities. The following table lists each Event Management KPI and the question it is trying to answer:

Table 17: Event Management KPI Performance Points

KPI	Question Being Answered
Incident Customer Impact Ratio	How many incidents do we resolve before they impact customers?
Percent Of Incidents Triggered By An Event	How often do we detect incidents before our customers do?
Event Noise Ratio	How often do we generate events that do not require a response?
Event Automation Ratio	How often do we handle events without human intervention?
Event Monitoring Coverage Ratio	How much of our infrastructure is monitored for events?
Average Event Delivery Time (Seconds)	Are events being communicated efficiently?
Event Management Tooling Support Level	How well does our current tool set support Event Management activities?
Event Management Process Maturity	How good are our Event Management practices?

Critical Success Factors (CSFs)

The table below lists suggested Critical Success Factors for Event Management. The KPI references listed in the right column indicate which KPIs are used as input for the associated CSF.

Table 18: Event Management CSFs

CSF	KPI
Detect And Resolve Incidents Before They Impact Customers	1,2
Provide Data For Service Assurance And Improvement	3,5,7,8
Capture Changes Of State For Assets And Services	2,3,7,8
Ensure Events Timely Communicated To IT Support Functions	3,4,6,7

Chapter

8

Access Management Metrics

Operational Metrics

The following table lists suggested Access Management Operational Metrics.

Table 19: Access Management Operational Metrics

XREF	METRIC
A	Total Number Of Incidents
B	Total Number of Security Access Incidents
C	Number Of Security Access Controls
D	Number Of Security Access Controls Failing Audits
E	Number Of Requests For Security Access Changes
F	Number Of Incidents Escalated To Security Teams
G	Average Notification Time For Access Violations (Minutes)
H	Total Labor Hours Spent Resolving Incidents (Non-Service Desk)
I	Number of Requests For Password Resets

XREF	METRIC
J	Number of Requests For Password Resets Fulfilled Automatically
K	Access Management Tooling Support Level
L	Access Management Process Maturity

Suggested sources for Access Management Operational Metrics can be found in places such as:

- ✓ Incident Management System
- ✓ Security Controls Policies
- ✓ Security Audit Findings
- ✓ Request Management System
- ✓ Labor Reports
- ✓ CMMI Tool Ratings or Tool Surveys
- ✓ Process Assessment Results

Key Performance Indicators (KPIs)

The following table lists suggested KPIs and how they are calculated from the Operational Metrics listed previously.

Table 20: Access Management KPIs

XREF	KPI	CALCULATION
1	Security Access Incident Rate	B/A
2	Security Access Audit Success Rate	D/C
3	Security Access Rights Accuracy Level	E
4	Security Escalation Rate For Access Violations	F/B
5	Average Access Violation Notification Response Time (Minutes)	G
6	Password Reset Automation Rate	J/I
7	Access Management Tooling Support Level	K
8	Access Management Process Maturity	L

Why These Metrics (KPIs) Matter

The KPIs described earlier are critical to managing and controlling Access Management activities. The following table lists each Access Management KPI and the question it is trying to answer:

Table 21: Access Management KPI Performance Points

KPI	Question Being Answered
Security Access Incident Rate	How many security incidents did we experience within our infrastructure?
Security Access Audit Success Rate	Are only those authorized allowed access to services?
Security Access Rights Accuracy Level	Are access rights accurate and in line with business needs?
Security Escalation Rate For Access Violations	How often do access incidents require a security escalation?
Average Access Violation Notification Response Time (Minutes)	How quickly do we notify security about access violations?
Password Reset Automation Rate	How efficient are we at providing self-service password resets?
Access Management Tooling Support Level	How well does our current tool set support Access Management activities?
Access Management Process Maturity	How good are our Access Management practices?

Critical Success Factors (CSFs)

The table below lists suggested Critical Success Factors for Access Management. The KPI references listed in the right column indicate which KPIs are used as input for the associated CSF.

Table 22: Access Management CSFs

CSF	KPI
Ensure confidentiality, integrity and availability of services	1,2,3,7,8
Provide access to services per agreed information security policies	2,3,5
Provide timely notices of improper access or abuse of services	5
Effectively and efficiently administer access to IT services	2,3,5,6,7,8

Chapter

9

Service Desk Metrics

Operational Metrics

The following table lists suggested Service Desk Operational Metrics.

Table 23: Service Desk Operational Metrics

XREF	METRIC
A	Total Number Of Calls To Service Desk
B	Average Call Duration (Minutes)
C	Average Call Waiting (Minutes)
D	Service Desk Tooling Support Level
E	Number of Calls Escalated
F	Number of Calls Abandoned
G	Available Call Agent Labor Hours
H	Total Service Desk Available Hours
I	Total Service Desk Unavailable Hours

Suggested sources for Service Desk Operational Metrics can be found in places such as:

- ✓ Automatic Call Distribution (ACD) Reports
- ✓ Incident Management System Reports
- ✓ Service Level Agreements (SLAs)
- ✓ Tool Assessment Audit Findings

Key Performance Indicators (KPIs)

The following table lists suggested KPIs and how they are calculated from the Operational Metrics listed previously.

Table 24: Service Desk KPIs

XREF	KPI	CALCULATION
1	Service Desk First Call Resolution Rate	1-((E+F)/A)
2	Average Call Duration (Minutes)	B
3	Service Desk Tooling Support Level	D
4	Call Agent Utilization	((B*A)/60)/G
5	Call Abandon Rate	F/A
6	Call Duration Experience	B+C
7	Call Waiting Rate	C/(C+B)
8	Service Desk Service Availability	1-(I/H)

Why These Metrics (KPIs) Matter

The KPIs described earlier are critical to managing and controlling Service Desk activities. The following table lists each Service Desk KPI and the question it is trying to answer:

Table 25: Service Desk KPI Performance Points

KPI	Question Being Answered
Service Desk First Call Resolution Rate	How many calls are resolved at the Service Desk on the first call?
Average Call Duration (Minutes)	How long is the average customer call experience?
Service Desk Tooling Support Level	How well does our current tool set support Service Desk activities?
Call Agent Utilization	Do we have enough resources to handle calls?
Call Abandon Rate	What percent of callers hang up before getting service?
Call Duration Experience	What is the total time taken to service callers?
Call Waiting Rate	What percentage of total caller time is spent waiting?
Service Desk Service Availability	Is the Service Desk available when needed?

Critical Success Factors (CSFs)

The table below lists suggested Critical Success Factors for the Service Desk function. The KPI references listed in the right column indicate which KPIs are used as input for the associated CSF.

Table 26: Service Desk CSFs

CSF	KPI
Resolve Customer Issues And Problems At First Call	1,3
Maintain Customer Productivity	1,3,6
Provide A Positive Customer Call Experience	3,5,6,7,8
Provide Effective Support For Customer Calls	3,4

Chapter
10

Change Management Metrics

Operational Metrics

The following table lists suggested Change Management Operational Metrics.

Table 27: Change Management Operational Metrics

XREF	METRIC
A	Total Changes In Pipeline
B	Total Changes Implemented
C	Number Of Failed Changes
D	Number of Emergency Changes
E	Number of Unauthorized Changes Detected
F	Number of Changes Rescheduled
G	Average Process Time Per Change (Days)
H	Number of Changes Resulting In Incidents
I	Change Management Tooling Support Level
J	Change Management Process Maturity
K	Total Available Labor Hours To Coordinate (Not Implement) Changes
L	Total Labor Hours Spent Coordinating (Not Implementing) Changes

Suggested sources for Change Management Operational Metrics can be found in places such as:

- ✓ Change Management System Reports
- ✓ Incident Management System Reports
- ✓ Labor or Other HR Reports
- ✓ Process and Tool Assessment Audit Findings
- ✓ Observations of Incidents or CMDB/Asset Reports to Detect Unauthorized Changes

Key Performance Indicators (KPIs)

The following table lists suggested KPIs and how they are calculated from the Operational Metrics listed previously.

Table 28: Change Management KPIs

XREF	KPI	CALCULATION
1	Change Efficiency Rate	B/A
2	Change Success Rate	1-(C/B)
3	Emergency Change Rate	D/A
4	Change Reschedule Rate	F/A
5	Average Process Time Per Change (Days)	G
6	Unauthorized Change Rate	E/B
7	Change Incident Rate	H/B
8	Change Labor Workforce Utilization	L/K
9	Change Management Tooling Support Level	I
10	Change Management Process Maturity	J

Why These Metrics (KPIs) Matter

The KPIs described earlier are critical to managing and controlling Change Management activities. The following table lists each Change Management KPI and the question it is trying to answer:

Table 29: Change Management KPI Performance Points

KPI	Question Being Answered
Change Efficiency Rate	How efficient are we at handling changes?
Change Success Rate	How effective are we at handling changes?
Emergency Change Rate	What percentage of changes were emergencies?
Change Reschedule Rate	How well do we implement changes on schedule?
Average Process Time Per Change (Days)	How long does the average change take?
Unauthorized Change Rate	What percentage of changes bypassed the Change process?
Change Incident Rate	What percentage of changes caused incidents?
Change Labor Workforce Utilization	How much available labor capacity was used to handle and coordinate changes?
Change Management Tooling Support Level	How well does our current tool set support Change Management activities?
Change Management Process Maturity	How well do we execute our Change Management practices?

Critical Success Factors (CSFs)

The table below lists suggested Critical Success Factors for Change Management. The KPI references listed in the right column indicate which KPIs are used as input for the associated CSF.

Table 30: Change Management CSFs

CSF	KPI
Protect Services When Making Changes	3,6,7
Make Changes Quickly And Accurately In Line With Business Needs	4,5,7,8,9
Make Changes Efficiently And Effectively	1,2,5,9
Utilize A Repeatable Process For Handling Changes	3,6,9,10

Chapter

11

Release Management Metrics

Operational Metrics

The following table lists suggested Release Management Operational Metrics.

Table 31: Release Management Operational Metrics

XREF	METRIC
A	Total Releases In Pipeline
B	Total Releases Implemented
C	Number Of Failed Releases
D	Number of Releases Rescheduled
E	Average Process Time Per Release
F	Number of Releases Resulting In Incidents
G	Release Management Tooling Support Level
H	Release Management Process Maturity
I	Total Available Labor Hours Implementing Releases
J	Total Used Labor Hours Implementing Releases
K	Number of Release Errors In Production
L	Number of Releases Withdrawn

Suggested sources for Release Management Operational Metrics can be found in places such as:

- ✓ Release Management System Reports
- ✓ Incident Management System Reports
- ✓ Labor or Other HR Reports
- ✓ Process and Tool Assessment Audit Findings
- ✓ Project Management Reports
- ✓ Project Status Reports

Key Performance Indicators (KPIs)

The following table lists suggested KPIs and how they are calculated from the Operational Metrics listed previously.

Table 32: Release Management KPIs

XREF	KPI	CALCULATION
1	Release Efficiency Rate	B/A
2	Release Success Rate	1-(C/B)
3	Release Reschedule Rate	D/A
4	Release Defect Rate	F/B
5	Release Labor Utilization	J/I
6	Release Management Tooling Support Level	G
7	Release Management Process Maturity Level	H
8	Number of Known Release Errors In Production	K
9	Release Withdraw Rate	L/A
10	Release Labor Waste Rate	(C+F+L)/A

Why These Metrics (KPIs) Matter

The KPIs described earlier are critical to managing and controlling Release Management activities. The following table lists each Release Management KPI and the question it is trying to answer:

Table 33: Release Management KPI Performance Points

KPI	Question Being Answered
Release Efficiency Rate	How efficient are we at handling releases?
Release Success Rate	How successful are we at implementing releases?
Release Reschedule Rate	How well do we implement releases on schedule?
Release Defect Rate	What percentage of releases caused incidents?
Release Labor Utilization	How much labor capacity was used to handling releases?
Release Management Tooling Support Level	How well does our current tool set support Release Management activities?
Release Management Process Maturity Level	How well do we execute our Release Management practices?
Number of Known Release Errors In Production	How good is the quality of releases put into production?
Release Withdraw Rate	What percent of releases never go into production?
Release Labor Waste Rate	What percent of release labor is wasted?

Critical Success Factors (CSFs)

The table below lists suggested Critical Success Factors for Release Management. The KPI references listed in the right column indicate which KPIs are used as input for the associated CSF.

Table 34: Release Management CSFs

CSF	KPI
Provide Repeatable Process For Rolling Out Releases	1,7
Implement High Quality Releases	2,4,8,9
Implement Releases Efficiently And Effectively	1,2,3,4,5,7,8,9,10
Implement Releases At Acceptable Costs	1,3,4,5,7,9,10

Chapter

12

Configuration Management Metrics

Operational Metrics

The following table lists suggested Configuration Management Operational Metrics.

Table 35: Configuration Management Operational Metrics

XREF	METRIC
A	Total Number of CIs in CMDB
B	Number of CIs Audited
C	Number of CI Errors Discovered
D	Configuration Management Tooling Support Level
E	Configuration Management Process Maturity
F	Number of CI Changes
G	Number of CI Changes Without Corresponding RFC
H	Number of Incidents Related To Inaccurate CI Information
I	Number of Change Failures Related To Inaccurate CI Information

XREF	METRIC
J	Number of Services Operating With Incomplete CI Information
K	Number Of Services In Service Catalog
L	Number of CIs Without Assigned Ownership

Suggested sources for Configuration Management Operational Metrics can be found in places such as:

- ✓ CMDB Audit and Status Accounting Reports
- ✓ Incident Management System Reports
- ✓ Change Management System Reports
- ✓ Process and Tool Assessment Audit Findings
- ✓ Service Catalog Listings
- ✓ Auto-Discovery Reports

Key Performance Indicators (KPIs)

The following table lists suggested KPIs and how they are calculated from the Operational Metrics listed previously.

Table 36: Configuration Management KPIs

XREF	KPI	CALCULATION
1	CMDB Accuracy Ratio	1-(C/A)
2	Number of Incidents Related To Inaccurate CI Information	H
3	Number of Change Failures Related To Inaccurate CI Information	I
4	Configuration Management Tooling Support Level	D
5	Configuration Management Process Maturity	E
6	CMDB Completeness Ratio	1-(J/K)
7	CI Ownership Rate	1-(L/A)

Why These Metrics (KPIs) Matter

The KPIs described earlier are critical to managing and controlling Configuration Management activities. The following table lists each Configuration Management KPI and the question it is trying to answer:

Table 37: Configuration Management KPI Performance Points

KPI	Question Being Answered
CMDB Accuracy Ratio	How accurate is information in the CMDB?
Number of Incidents Related To Inaccurate CI Information	How many incidents were related to inaccurate configuration information?
Number of Change Failures Related To Inaccurate CI Information	How many changes failed due to inaccurate configuration information?
Configuration Management Tooling Support Level	How well does our current tool set support Configuration Management activities?
Configuration Management Process Maturity	How well do we execute our Configuration Management practices?
CMDB Completeness Ratio	How complete is our configuration information?
CI Ownership Rate	How much of our infrastructure has no assigned ownership?

Critical Success Factors (CSFs)

The table below lists suggested Critical Success Factors for Configuration Management. The KPI references listed in the right column indicate which KPIs are used as input for the associated CSF.

Table 38: Configuration Management CSFs

CSF	KPI
Control Information About The IT Infrastructure	5,6,7
Support Delivery Of Quality IT Services	2
Maintain Accurate Configuration Information	1,3
Support IT Service Management Processes	3,4,5

Chapter

13

Service Level Management Metrics

Operational Metrics

The following table lists suggested Service Level Management Operational Metrics.

Table 39: Service Level Management Operational Metrics

XREF	METRIC
A	Overall Customer Satisfaction Rating
B	Number Of IT Business Facing Services Delivered To Customers
C	Number Of IT Business Facing Services Without SLAs
D	Number Of IT Support Services SLAs (OLAs)
E	Number Of IT Support Services Without SLAs or OLAs
F	Number Of Services Delivered By Vendors
G	Number Of Vendor Services Without Agreed Service Targets
H	Total Service Penalties Paid
I	Total Number Of SLA Service Targets

XREF	METRIC
J	Total Number Of SLA Service Targets Breached
K	Number Of SLAs Operating Without Service Ownership
L	Service Level Management Tooling Support Level
M	Service Level Management Process Maturity

Suggested sources for Service Level Management Operational Metrics can be found in places such as:

- ✓ Customer Survey Results
- ✓ Service Catalog Listings
- ✓ Procurement Contract Files
- ✓ Accounts Payable Reports
- ✓ Service Level Agreements
- ✓ Service Level Reporting
- ✓ Service Level Management Database Reports
- ✓ Tool And Process Assessment Audit Findings

Key Performance Indicators (KPIs)

The following table lists suggested KPIs and how they are calculated from the Operational Metrics listed previously.

Table 40: Service Level Management KPIs

XREF	KPI	CALCULATION
1	Overall Customer Satisfaction Rating	A
2	SLA Coverage Rate	1-(C/B)
3	OLA Coverage Rate	1-(E/D)
4	Percent Of Vendor Services Delivered Without Agreed Service Targets	G/F
5	Total Service Penalties Paid	H
6	Percent Of SLA Service Targets Adhered To	1-(J/I)
7	Percent Of SLAs With Responsible Service Owners	1-(K/B)
9	Service Level Management Tooling Support Level	L
10	Service Level Management Process Maturity	M

Why These Metrics (KPIs) Matter

The KPIs described earlier are critical to managing and controlling Service Level Management activities. The following table lists each Service Level Management KPI and the question it is trying to answer:

Table 41: Service Level Management KPI Performance Points

KPI	Question Being Answered
Overall Customer Satisfaction Rating	How do customers perceive the quality of the services we are delivering?
SLA Coverage Rate	What percent of the services we deliver have formally been agreed to?
OLA Coverage Rate	What percent of our internal supporting services operate with formal agreements?
Percent Of Vendor Services Delivered Without Agreed Service Targets	What percent of our supporting services are delivered without agreed service targets?
Total Service Penalties Paid	How much are we paying in service penalties?
Percent Of SLA Service Targets Adhered To	How well have we met our SLA service targets?
Percent Of SLAs With Responsible Service Owners	What percent of our delivered services do not have assigned service owners?
Service Level Management Tooling Support Level	How well does our current tool set support Service Level Management activities?
Service Level Management Process Maturity	How well do we execute our Service Level Management Processes?

Critical Success Factors (CSFs)

The table below lists suggested Critical Success Factors for Service Level Management. The KPI references listed in the right column indicate which KPIs are used as input for the associated CSF.

Table 42: Service Level Management CSFs

CSF	KPI
Deliver IT Services As Agreed To By Customers And The Business	2,3,4,5,6
Support The Business/User Interface	1,7,8,9
Provide Services At Acceptable Cost	5
Manage Quality Of IT Services In Line With Business Requirements	1,5,6,8,9

Chapter

14

Availability Management Metrics

Operational Metrics

The following table lists suggested Availability Management Operational Metrics.

Table 43: Availability Management Operational Metrics

XREF	METRIC
A	Total Unplanned Expenses Related To Availability
B	Total Number of Incidents
C	Total Number Of Customer Impacting Incidents
D	Total Available Minutes For All Services Delivered
E	Total Unavailable Minutes For All Services Delivered
F	Availability Management Tooling Support Level
G	Availability Management Process Maturity Level

XREF	METRIC
H	Total Number of Service Targets From Internal Suppliers
I	Total Number of Service Targets From Vendor Suppliers
J	Number of Internal Supplier Targets Missed
K	Number of Vendor Supplier Targets Missed
L	Number of Security Related Incidents
M	Number of HW/SW/Networking CIs
N	Number of HW/SW/Networking CIs Not Supported By Vendors
O	Number Of Services In Service Catalog
P	Number of Services Not Covered By An Active Availability Plan
Q	Number of Services Without Availability Review Last 3 Months

Suggested sources for Availability Management Operational Metrics can be found in places such as:

- ✓ Incident Management System
- ✓ Problem Management System
- ✓ Service Catalog Listings
- ✓ Service Level Agreements
- ✓ Operational Level Agreements
- ✓ Underpinning Contracts
- ✓ Configuration Management Database Reports
- ✓ Existing Availability Plans
- ✓ Service Level Reporting
- ✓ Service Level Management Database Reports
- ✓ Tool And Process Assessment Audit Findings

Key Performance Indicators (KPIs)

The following table lists suggested KPIs and how they are calculated from the Operational Metrics listed previously.

Table 44: Availability Management KPIs

XREF	KPI	CALCULATION
1	Total Unplanned Expenses Related To Availability	A
2	Availability Resilience Index	1-(C/B)
3	Average Service Reliability Index	1-(E/D)
4	Availability Management Tooling Support Level	F
5	Availability Management Process Maturity Level	G
6	Average Internal Supplier Service Reliability Index	1-(J/H)
7	Average Vendor Supplier Service Reliability Index	1-(K/I)
8	Security Vulnerability Index	L/B
9	Serviceability Index	N/M
10	Availability Risk Index	P/O
11	Continuous Availability Improvement Index	1-(Q/O)

Why These Metrics (KPIs) Matter

The KPIs described earlier are critical to managing and controlling Availability Management activities. The following table lists each Availability Management KPI and the question it is trying to answer:

Table 45: Availability Management KPI Performance Points

KPI	Question Being Answered
Total Unplanned Expenses Related To Availability	How much unplanned costs were spent on maintaining needed availability?
Availability Resilience Index	How resilient is our infrastructure towards protecting services?
Average Service Reliability Index	How reliable are the services we deliver?
Availability Management Tooling Support Level	How well does our current tool set support Availability Management activities?
Availability Management Process Maturity Level	How well do we execute our Availability Management practices?
Average Internal Supplier Service Reliability Index	How reliably are internal suppliers supporting our services?
Average Vendor Supplier Service Reliability Index	How reliably are vendor suppliers supporting our services?
Security Vulnerability Index	How vulnerable are we to security threats?
Serviceability Index	How much of our physical infrastructure is supported by vendors?

KPI	Question Being Answered
Availability Risk Index	What percent of our services are delivered without addressing availability?
Continuous Availability Improvement Index	How well do we proactively look at improving service availability?

Critical Success Factors (CSFs)

The table below lists suggested Critical Success Factors for Availability Management. The KPI references listed in the right column indicate which KPIs are used as input for the associated CSF.

Table 46: Availability Management CSFs

CSF	KPI
Provide Services With Availability To Match Business Need	2,8
Demonstrate Cost-Effectiveness Through Availability Planning	1,2,3,4,5,6,7,8,9,10,11
Continually Improve Availability Of Delivered Services	11
Effectively Manage Availability Risks	1,2,3,4,5,6,7,8,9

Chapter

15

Capacity Management Metrics

Operational Metrics

The following table lists suggested Capacity Management Operational Metrics.

Table 47: Capacity Management Operational Metrics

XREF	METRIC
A	Total Expenses For Unplanned Capacity
B	Number of IT Resource Forecasts
C	Number of IT Service Forecasts
D	Number of IT Business Forecasts
E	Number of Missed IT Resource Forecasts
F	Number of Missed IT Service Forecasts
G	Number of Missed IT Business Forecasts
H	Number of Incidents Caused By Inadequate Capacity
I	Total Actual IT Costs For Hardware, Software and Network
J	Capacity Management Tooling Support Level
K	Capacity Management Process Maturity Level

Suggested sources for Capacity Management Operational Metrics can be found in places such as:

- ✓ Capacity Plans and Reports
- ✓ IT Financial Accounting and Budget Reports
- ✓ Incident Management System Reports
- ✓ Procurement Reports
- ✓ Tool And Process Assessment Audit Findings

Key Performance Indicators (KPIs)

The following table lists suggested KPIs and how they are calculated from the Operational Metrics listed previously.

Table 48: Capacity Management KPIs

XREF	KPI	CALCULATION
1	Total Expenses For Unplanned Capacity	A
2	Percent of IT Costs For Unplanned Capacity Expenses	A/I
3	IT Resource Forecast Accuracy Ratio	1-(E/B)
4	IT Service Forecast Accuracy Ratio	1-(F/C)
5	IT Business Forecast Accuracy Ratio	1-(G/D)
6	Number of Incidents Caused By Inadequate Capacity	H
7	Capacity Management Tooling Support Level	J
8	Capacity Management Process Maturity Level	K

Why These Metrics (KPIs) Matter

The KPIs described earlier are critical to managing and controlling Capacity Management activities. The following table lists each Capacity Management KPI and the question it is trying to answer:

Table 49: Capacity Management KPI Performance Points

KPI	Question Being Answered
Total Expenses For Unplanned Capacity	How much did unplanned capacity cost us for HW/SW/Network Components?
Percent of IT Costs For Unplanned Capacity Expenses	What percent of our actual HW/SW/Network costs were for unplanned capacity?
IT Resource Forecast Accuracy Ratio	How accurate are we in forecasting IT Needed Resources?
IT Service Forecast Accuracy Ratio	How accurate are we in predicting workload volumes for services?
IT Business Forecast Accuracy Ratio	How accurate are we in anticipating business growth and changes?
Number of Incidents Caused By Inadequate Capacity	How many incidents were caused related to capacity?
Capacity Management Tooling Support Level	How well does our current tool set support Capacity Management activities?
Capacity Management Process Maturity Level	How well do we execute our Capacity Management practices?

Critical Success Factors (CSFs)

The table below lists suggested Critical Success Factors for Capacity Management. The KPI references listed in the right column indicate which KPIs are used as input for the associated CSF.

Table 50: Capacity Management CSFs

CSF	KPI
Provide Accurate Capacity Forecasts	3,4,5
Provide Services With Appropriate Capacity To Match Business Need	6,8
Protect Services From Capacity Related Incidents	6
Demonstrate Cost-Effectiveness Through Accurate Capacity Planning	1,2,7

Chapter

16

IT Service Continuity Management Metrics

Operational Metrics

The following table lists suggested IT Service Continuity Management Operational Metrics.

Table 51: Service Continuity Operational Metrics

XREF	METRIC
A	Number Of Services In Service Catalog
B	Number of Services Covered By IT Service Continuity Plans
C	Number of Service Continuity Plan Audit Failures
D	Mean Time (Months) Between Continuity Tests For Each Service
E	Total Planned Labor Hours For IT Service Continuity Activities
F	Total Used Labor Hours For IT Service Continuity Activities
G	Number of IT Services Tested For Service Continuity

XREF	METRIC
H	Mean Time (Months) Between Continuity Plan Audits For Each Service
I	Total Planned Costs For ITSCM Activities
J	Total Unplanned Costs For ITSCM Activities
K	Number of Business Continuity Issues Logged Against ITSCM Plans
L	IT Service Continuity Tooling Support Level
M	IT Service Continuity Management Process Maturity
N	Number of Services With Test Failures
O	Total Number Of Services Needed To Support ITSCM Plans
P	Number Of Required Internal Support Services Without An OLA
Q	Number Of Required External Support Services Without Formal Agreements

Suggested sources for IT Service Continuity Management Operational Metrics can be found in places such as:

- ✓ IT Service Continuity Plans
- ✓ Service Catalog Listings
- ✓ Service Continuity Test Plans
- ✓ Service Continuity Test Results
- ✓ HR and Labor Reports
- ✓ IT Financial Accounting and Budget Reports
- ✓ Operational Level Agreements (OLAs)
- ✓ Underpinning Contracts (UCs)
- ✓ Tool And Process Assessment Audit Findings

Key Performance Indicators (KPIs)

The following table lists suggested KPIs and how they are calculated from the Operational Metrics listed previously.

Table 52: Service Continuity KPIs

XREF	KPI	CALCULATION
1	ITSCM Coverage Ratio	B/A
2	Number of Service Continuity Plan Audit Failures	C
3	Mean Time (Months) Between Continuity Tests For Each Service	D
4	ITSCM Labor Utilization	F/E
5	Testing Completeness Ratio	G/A
6	ITSCM Cost Rate	J/I
7	IT Service Continuity Tooling Support Level	L
8	IT Service Continuity Management Process Maturity	M
9	IT Service Continuity Recovery Confidence Rate	1-(N/A)
10	Number of Business Continuity Issues Logged Against ITSCM Plans	K
11	Mean Time Between Continuity Plan Audits For Each Service	H
12	ITSCM Support Service Coverage Ratio	1-((P+Q)/O)

Why These Metrics (KPIs) Matter

The KPIs described earlier are critical to managing and controlling IT Service Continuity Management activities. The following table lists each IT Service Continuity Management KPI and the question it is trying to answer:

Table 53: Service Continuity KPI Performance Points

KPI	Question Being Answered
ITSCM Coverage Ratio	How much of our IT Services are covered under the ITSCM Plan?
Number of Service Continuity Plan Audit Failures	How reliable is our ITSCM Plan?
Mean Time (Months) Between Continuity Tests For Each Service	How confident are we that enough testing has been done?
ITSCM Labor Utilization	How much of our planned labor capacity was used for ITSCM activities?
Testing Completeness Ratio	How complete is our testing?
ITSCM Cost Rate	How well did we plan for ITSCM Costs?
IT Service Continuity Tooling Support Level	How well does our current tool set support ITSCM activities?
IT Service Continuity Management Process Maturity	How well do we execute our ITSCM practices?
IT Service Continuity Recovery Confidence Rate	How much confidence do we have that we can recover needed services?

KPI	Question Being Answered
Number of Business Continuity Issues Logged Against ITSCM Plans	How aligned is our ITSCM plan with the Business Continuity Plan?
Mean Time Between Continuity Plan Audits For Each Service	How aligned is our ITSCM plan with our current infrastructure?
ITSCM Support Service Coverage Ratio	Do we have all necessary agreements with recovery support suppliers?

Critical Success Factors (CSFs)

The table below lists suggested Critical Success Factors for IT Service Continuity Management. The KPI references listed in the right column indicate which KPIs are used as input for the associated CSF.

Table 54: Service Continuity CSFs

CSF	KPI
Recover Services From Major Disruptions Within Business Timeframes	2,5,8,9,10,12
Ensure All Required Services Can Be Recovered From Major Disruptions	1,8,12
Maintain Viability Of IT Service Continuity Plans	3,8,10,11
Provide Service Continuity At Acceptable Costs	4,6,7

Chapter

17

IT Financial Management Metrics

Operational Metrics

The following table lists suggested IT Financial Management Operational Metrics.

Table 55: Financial Management Operational Metrics

XREF	METRIC
A	Total Infrastructure Budget
B	Total To-Date Planned Budget Costs
C	Total To-Date Actual Budget Costs
D	Financial Management Tooling Support Level
E	Financial Management Process Maturity
F	Total Actual Chargeback Revenue
G	Total Planned Chargeback Revenue
H	Number Of Financial Reports/Forecasts Delivered Late
I	Number Of Changes To Charging Algorithms

Suggested sources for IT Financial Management Operational Metrics can be found in places such as:

- ✓ Chargeback Reports
- ✓ Capacity Plans
- ✓ IT Financial Accounting and Budget Reports
- ✓ Financial Reporting Production Schedules
- ✓ Tool And Process Assessment Audit Findings

Key Performance Indicators (KPIs)

The following table lists suggested KPIs and how they are calculated from the Operational Metrics listed previously.

Table 56: Financial Management KPIs

XREF	KPI	CALCULATION
1	Unplanned Cost Variance	B-A
2	Cost Performance Index	B/C
3	Estimated Year-End Costs	A/(B/C)
4	Variance At Budget Year-End	(A/(B/C))-A
5	Financial Management Tooling Support Level	D
6	Financial Management Process Maturity	E
7	IT Revenue Variance	H-G
8	Number Of Financial Reports/Forecasts Delivered Late	H
9	Number Of Changes To Charging Algorithms	I

Why These Metrics (KPIs) Matter

The KPIs described earlier are critical to managing and controlling IT Financial Management activities. The following table lists each IT Financial Management KPI and the question it is trying to answer:

Table 57: Financial Management KPI Performance Points

KPI	Question Being Answered
Unplanned Cost Variance	Are we operating within planned costs?
Cost Performance Index	What is the ratio of expected costs to actual costs?
Estimated Year-End Costs	How much do we think we will actually spend by end of year?
Variance At Budget Year-End	How much more cost do we anticipate is needed by year-end?
Financial Management Tooling Support Level	How well does our current tool set support Financial Management activities?
Financial Management Process Maturity	How well do we execute our Financial Management practices?
IT Revenue Variance	How much chargeback revenue did we make compared to our plan?
Number Of Financial Reports/Forecasts Delivered Late	Are we delivering financial information to the business on time?
Number Of Changes To Charging Algorithms	Is there satisfaction with how IT charges for its services?

Critical Success Factors (CSFs)

The table below lists suggested Critical Success Factors for IT Financial Management. The KPI references listed in the right column indicate which KPIs are used as input for the associated CSF.

Table 58: Financial Management CSFs

CSF	KPI
Provide Effective Stewardship Of IT Finances	1,2,3,4,6,8
Maintain Overall Effectiveness Of The IT Financial Management Process	5,6,8
Ensure Customers Satisfied With Costs And Charges For Services	9
Recapture IT Costs Through Chargeback For Delivery Of It Services	5,7

Chapter

18

IT Workforce Management Metrics

Operational Metrics

The following table lists suggested IT Workforce Management Operational Metrics.

Table 59: Workforce Management Operational Metrics

XREF	METRIC
A	Total Number Of IT Service Delivery And Support Staff
B	Number Of IT Staff With Certifications In IT Service Management
C	Number Of IT Staff Actively Participating In Industry Quality Organizations
D	IT Employee Satisfaction Survey Rating
E	IT Staff Turnover Rate
F	Non-Value Labor Rating

Suggested sources for IT Workforce Management Operational Metrics can be found in places such as:

- ✓ Labor and HR Reports
- ✓ Performance Reviews and Appraisals
- ✓ Employee Satisfaction Survey Results
- ✓ Employee Time Reporting Reports
- ✓ Employee Time Usage Surveys or Observations

Key Performance Indicators (KPIs)

The following table lists suggested KPIs and how they are calculated from the Operational Metrics listed previously.

Table 60: Workforce Management KPIs

XREF	KPI	CALCULATION
1	IT Staff Service Management Certification Rate	B/A
2	IT Staff Participation In External Industry Quality Organizations	C/A
3	IT Employee Satisfaction Survey Rating	D
4	IT Staff Turnover Rate	E
5	Non-Value Labor Rating	F

Why These Metrics (KPIs) Matter

The KPIs described earlier are critical to managing and controlling IT Workforce Management activities. The following table lists each IT Workforce Management KPI and the question it is trying to answer:

Table 61: Workforce Management KPI Performance Points

KPI	Question Being Answered
IT Staff Service Management Certification Rate	How skilled is our IT staff in IT Service Management?
IT Staff Participation In External Industry Quality Organizations	How much of our IT staff participate in outside industry quality organizations?
IT Employee Satisfaction Survey Rating	How satisfied is our IT Staff with the work they do?
IT Staff Turnover Rate	How well do we retain IT staff, skills and experience?
Non-Value Labor Rating	How much IT labor is spent on non-value activities?

Critical Success Factors (CSFs)

The table below lists suggested Critical Success Factors for IT Workforce Management. The KPI references listed in the right column indicate which KPIs are used as input for the associated CSF.

Table 62: Workforce Management CSFs

CSF	KPI
Maximize IT Staff Productivity	5
Provide High Level Of Staff Skills In IT Service Quality	1,2
Maintain Positive Staff Morale	3,4

Chapter
19

Measuring Services

Key Metrics for Measuring Service Quality

This chapter provides an inventory of Critical Success Factors (CSFs) and Key Performance Indicators (KPIs) that can be used to report on services, their quality and value to the business organization. A balanced scorecard approach has been taken to categorize each key performance indicator. A Service Metrics spreadsheet tool has also been provided that will allow you to set action thresholds and report on the services that you provide.

Customer Service Metrics

These are metrics around customer perception of the service. They include measurements such as customer satisfaction, customer loyalty, and market share.

Table 63: Service Metrics — Customer

CSF	DESCRIPTION	KPI
Market Share	Percent of the market served by the service	Percent of customer population using the service
Customer Satisfaction	How well service meets customer expectations	Satisfaction survey on a 1 to 10 scale where 10 is extremely satisfied
Adherence To Service Levels	How well service delivered within agreed service targets	Percent of all requests and incidents without service breaches
Customer Loyalty	How well customers return to use the service over and over	Customer growth or reduction

Operational Service Metrics

These are metrics that describe the internal operational efficiencies and effectiveness of activities and resources used to deliver the service.

Table 64: Service Metrics – Operational

CSF	DESCRIPTION	KPI
Service Availability	Percent of time service is available	Hours Available / All Planned Available Hours
Service Accuracy	Service reports and data meet expectations for accuracy	Number of failed accuracy audit observations / all audit observations
Service Performance	Service response times and overall performance meet targets	Percent of incidents reported with response/ performance issues
Supplier Performance	How well internal and external suppliers deliver their services within targets	Percent of all supplier requests and incidents without service breaches
Support Turnover	Average turnover rate for support staff	Average percent of turnover for support staff
Service Compliance	Service compliance with legal, regulatory and industry regulations	Percent of audit findings that failed compliancy
Service Flexibility	Service adapts easily and quickly to changing business needs	Percent of changes to service that were on time and completed successfully

Capability Service Metrics

These are metrics that describe the level of capabilities, skills, suppliers, capacity and performance of the service being delivered.

Table 65: Service Metrics – Capability

CSF	DESCRIPTION	KPI
Service Capacity	Service capacity is sized to meet customer current and future demands	Average utilization of supporting technologies and people resources
Supplier Performance	Supplier adherence service targets	Average percent attainment to agreed or contracted service targets
Market Performance	Alignment with other suppliers In the marketplace delivering similar service	Superior to competitors, par with competitors or below competitors
Service Recoverability	Ability to recover service in case of a major business disruption	Confidence level for full service recovery within required business targets
Support Skills	Internal staff skills and capabilities to support the service	Average percent of turnover for support staff
Technology Performance	Technology capability to automate and support the service	Strong, moderate or weak technology capabilities

CSF	DESCRIPTION	KPI
Supportability	Capability of suppliers to support the service throughout the marketplace	Service is fully supported, partially supported or not supported by suppliers
Market Growth	Ability to serve additional customer groups or expand into new markets	Strong, moderate or weak opportunities to expand marketplace footprint

Financial Service Metrics

These are metrics that identify the cost and revenue performance of the service being delivered.

Table 66: Service Metrics—Financial

CSF	DESCRIPTION	KPI
Cost Performance	Cost competitive—alignment to costs for similar service in the marketplace	Superior to competitors, par with competitors or below competitors
Revenue Performance	Aligned to business revenue goals for this service	Percent of revenue targets
Budget Performance	Adherence to planned costs and expenditures	Percent adherence to established budget

Installing the Service Metrics Tool

This tool is built as a Microsoft EXCEL Spreadsheet and included with the download files mentioned in Chapter 1. Simply download or copy the **Tooling Aid—ITSM Service Metrics Tool.xls** file to a desired folder on your PC. The PC itself should be running WINDOWS or other platform compatible with Microsoft Office.

It is recommended that you install the original version of the file and make changes only to copies of it. This will allow you to continually reuse and apply the original for each service that you wish to report on. For example:

1. Download and copy the original file to your PC or device as **Tooling Aid—ITSM Service Metrics Tool.xls**

2. Create a copy of the file for any given service and apply your metrics results. Store it as **MyService. xls** (for example).

3. Create additional copies for each service you wish to measure and report on. Make sure you change the name of each file to match the service that you are reporting against.

How to Use the ITSM Service Metrics Tool

The model is simple to use. It consists of an EXCEL Worksheet where you can input your KPI results. These are in the Metrics spreadsheet in the white colored boxes under the Actual Column. You can also enter thresholds and a weighting to indicate whether actions may be needed to address service issues. The remainder of the model, including the service results and dashboard gauge settings is automatically calculated for you.

To use the tool, go to the Metrics worksheet and execute the following steps:

1. Set the targets for each KPI
2. Set Warning threshold targets (indicates when yellow status occurs)
3. Set Action threshold targets (indicates when red status occurs)
4. Fill in Actuals column
5. Set weightings for KPIs as needed
6. Review your results

Interpreting the Service Metrics Results

Higher scores (RED) indicate presence of service issues while lower scores (GREEN) indicate services are being delivered successfully and providing value. Results that fall between these two (YELLOW) will indicate that some actions may be needed, or that services should be watched more closely to see if issues may need further attention.

Chapter

20

Alternatives If Few Metrics Available

Establishing a Minimal Metrics Program

As efforts begin to establish an ITSM Metrics Program, there may be discovery that sources for metrics are far and few between. Typical challenges may include:

- Many disparate data collection and reporting tools in place with poor ability to aggregate and summarize data
- No clear authoritative source for metrics
- No staffing priorities to collect analyze and report on metrics
- A lack of tools and automation to report on metrics

When faced with these challenges, there is a great temptation to just quit the effort and wait until management priorities change or tool funding occurs. This is a great mistake. Attempts should always be made to measure and manage with what you can whenever possible. The alternative is to do no measuring at all.

Not measuring at all is a much worse alternative than selecting and agreeing on some minimal measurements that can be used as a starting point. These measurements may not be as accurate as those supported by a fully funded tool and project effort, but they can still serve until such an effort can get started.

The main benefit of establishing a metrics program with minimal metrics is to establish a culture of measurement goals and focus on IT service quality. Management desire to obtain better accuracy and information in this environment can actually lead to Senior Executive funding and approval for what will really be needed.

A general approach for establishing a set of minimal metrics can be done as follows:

1) Select a small subset of measurements that are "representative" of the quality of service being delivered

2) Develop assumptions as to their accuracy level and how they will be used

3) Review and agree these with senior management

4) Report on these as if a full-fledged metrics program were in place

Some examples of the kinds of minimal metrics that could be collected in this kind of program include the following:

➢ Use Indicators
➢ Random or Scheduled Inspection Results
➢ Analogous Measures
➢ Programmed Measures
➢ Audit Results

Use Indicators

These are metrics that are based on some observable operational event for which an operating quality assumption is derived. Some examples might include:

- ✓ If the number of staff working on incidents throughout the day exceeds N, then it will be assumed that the acceptable service incident rate is too high

- ✓ If more than N incidents occur due to changes, then the Change Success Rate will be considered too low

- ✓ If the length of a service outage is N minutes, then the overall availability of that service will be assumed as

    1–(N/All planned availability minutes)

Random or Scheduled Inspection Results

These are metrics that represent observable events for which an assumption is made *that they apply to all similar events.* Some examples might include:

- ✓ If the utilization of network lines A, B and C are checked at 1:00pm on business days and found to be under N%, then it will be assumed that there is appropriate resource capacity to support services.

- ✓ If observed service restore times for the first 3 incidents of the week are X, Y and Z minutes, then it will be assumed that the Average Incident Restore time for all incidents is the average of those observations.

- ✓ If one remote network device has incurred N service outages it will be assumed that all remote network devices typically incur N service outages.

- ✓ Customer Satisfaction Survey results for one geographic location will represent similar results at all geographic locations.

Analogous Measures

These are observable metrics from which assumptions will be derived that other events have occurred. In general, this approach states something like: ". . . if we see *this*—it really means *that* is happening . . ." This is purely a set of assumptions and should be generally agreed to by IT management. Some examples might include:

✓ If the number of customer complaint calls exceeds N on any given day of the month, then it will be assumed that the Customer Satisfaction level is too low.

✓ If a remote network device has incurred a service outage for N minutes then it will be assumed that all IT services at that location were also out of service for N minutes.

✓ If it is determined that an Incident was caused by a change and there is no RFC (Request for Change) that represents the Change, then the Number of Unauthorized Changes will increase by 1.

✓ The number of calls to the Service Desk will represent the number of incidents that took place.

✓ If the IT headcount at one operational facility is greater than 20% of the average headcount across all facilities, then it will be assumed that there is a high level of non-value labor at that facility.

Programmed Measures

Creating Programmed Measures involves developing programs that will produce events or measures that will substitute for real events. It is a management decision as to whether the programs developed will produce accurate findings or create random or analogous measures. That decision will usually be based on how much time, labor and cost that the business is willing to invest to produce the measure. Some examples might include:

✓ Developing end-to-end response time observations by creating dummy online transactions that traverse the network similar to real transactions. The dummy transactions drop a log file time stamp when they kick off and return. The observed response time is the difference between the two time stamps.

✓ Using the same program and approach as described above, the dummy transaction will be scheduled to execute every N minutes. If no response is received within Y minutes, it will be assumed that the IT service has experienced an outage and the Y value will be subtracted from the overall availability measure for the IT service it represents.

✓ An online transaction is programmed to update a special log file with the count of key business activities selected by a user that will be used to report on Business Capacity Management drivers.

Audit Results

This consists of obtaining measures from periodic audit activities that are conducted for specific operational events. The results of the audit will then be applied to one or more KPI values. Some examples might include:

- ✓ Conducting a periodic Customer Satisfaction Survey and reporting the result as the Customer Satisfaction Level.

- ✓ Using a security penetration test to represent the quality of the availability of the IT infrastructure.

- ✓ Using an IT COBIT (Control Objectives for Information and Related Technology—a framework created by ISACA for IT management and governance) audit to assess the quality of the IT Change Management process.

Considerations for Using Minimal Metrics

The decision to use a Minimal Metrics Program should be based on the situation where metrics are really hard to obtain without a major project effort or tools investment. It should not be used as a quick shortcut substitute approach to avoid efforts to put in a quality Metrics solution.

Undertaking a minimal approach through the techniques described in this chapter will require a lot of creativity and management communications as to the measurement techniques being deployed. It means first identifying what the desired measurement set should be then identifying a set of techniques that will represent that measurement set.

At all times, the decision to undertake this kind of an effort will require constant management communications and a clear understanding and agreement to the techniques employed.

Chapter

21

Using the ITSM
Project Modeling Tool

ITSM Project Modeling Tool Overview

This model is a simple spreadsheet tool that can be used to predict the likelihood of success for an ITSM implementation or improvement effort. It is based on the DICE model that was originally developed by Harold L. Sirkin, Perry Keenan and Alan Jackson of the Boston Consulting Group who conducted a correlation study of 225 companies to determine the common denominators for successful organizational behavior change.

The DICE acronym represents the common denominators that were found. These are:

- **Duration**—length of time between project reviews

- **Integrity**—Extent to which the organization can rely upon the project team to execute the project successfully

- **Commitment**—Ensuring that appropriate levels of Senior Management and Stakeholder commitment are in place

- **Effort**—The estimated amount of time those making the change will have to spend over and above their day-to-day responsibilities.

These four elements are then combined into a Project Success Prediction Score. In their study, regression analysis revealed that the combination of the above listed factors that correlated the most closely with actual project outcomes doubled the weight that was given to the performance of the team and the commitment of Senior Management. In the Model, the Score Total is calculated from these factors using their formula.

For more specific information, readers are encouraged to read the white paper published by the authors. It can be found at:

Harvard Business Review: www.hbr.org
The Hard Side of Change Management
Harold L. Sirkin, Perry Keenan, Alan Jackson
Reprint: R0510G

Installing the Model

The model is built as a Microsoft EXCEL Spreadsheet and included with the download files described in Chapter 1. Simply download or copy the **Tooling Aid—ITSM Project Modeling Tool v3.xls** file to a desired folder on your PC. The PC itself should be running WINDOWS or other platform compatible with Microsoft Office.

It is recommended that you install the original version of the file and make changes only to copies of it. This will allow you to continually reuse the original to create baselines or future state models of your project environment. For example:

1. Download and copy the original file to your PC or device as **Tooling Aid—ITSM Project Modeling Tool v3.xls**

2. Create a baseline of your project current state environment by making a copy of this file, applying your metrics results and storing it as **MyProject Baseline.xls** (for example)

3. Create state models of future project improvement decisions (such as modeling the impact of getting increased Target Stakeholder buy-in or a more capable Project Manager) by creating copies of your baseline model and storing it with some relevant name (i.e. **MyProjectImprovements. xls** for example) and then apply changes to that baseline. In this way, you can create multiple versions of models based on different project improvement scenarios and compare their impact to the overall score results.

How to Use the ITSM Project Modeling Tool

The model is simple to use. It consists of an EXCEL Worksheet where you can input your project parameters. These are in the grey colored boxes. The remainder of the model, including the project prediction score and interpretation is automatically calculated for you.

Interpreting the Model Results

The model will calculate a Project Prediction score and then interpret this into the likelihood of success for your project into one of four areas. These are:

Win Zone
The project has a high likelihood that it will succeed.

Worry Zone
The project has a reasonable likelihood that it will succeed however there are some risk areas that should be watched closely. The risk areas are shown underneath the individual scores and color coded as Green, Yellow or Red. Any item coded as Red or Yellow with a score of 3 is a candidate area to watch closely.

Woe Zone
The project may not succeed unless certain risk areas are addressed. These are highlighted in yellow or red colors in the individual scores section of the model.

Disaster Zone
The project has a strong likelihood of NOT succeeding. The risk areas are highlighted in the individual scores section as described earlier.

Projects with Low Likelihood of Success

The purpose of using the model is to highlight project risk areas that need to be addressed. If the calculated project score puts the project in the Woe or Disaster Zone, all is not lost. The project team simply needs to address the risks that are indicated in Red. An approach for doing this could be done as follows:

1. Identify which risk areas have the highest scores

2. Model changes to those risks. For example: If the capabilities of the project team were initially scored as LOW, change this to HIGH and see the overall impact to the project. Did it the resulting Zone change for the better?

3. Determine which changes had the greatest positive impact to your project effort

4. Develop an action plan to achieve the changes that you modeled. For example: add more highly skilled staff to your project team (from the previous example)

5. Obtain approval for your recommended changes and implement them.

It is highly recommended that you re-model your project at each formal project review. This will allow you to identify risks early and mitigate them before they endanger your efforts. It may not be unusual to find that you may have started the effort with high stakeholder support initially, but that this has waned somewhat as the project proceeds over time. It is important to understand the impacts this could be having on your overall success.

Chapter

22

Using the ITSM Metrics Modeling Tool

ITSM Metrics Model Tool Overview

The ITSM Metrics Model is a simple spreadsheet tool that can be used for a variety of measurement and reporting purposes. The model can be used as:

    ✓ A starting point to identify key metrics that can be used to measure and monitor the health and state of your ITSM processes and activities

    ✓ Justifying an ITSM improvement initiative by modeling desired target future state improvements expected to occur

    ✓ A means for demonstrating the impacts and effects of current ITSM practices

✓ A means for modeling future business decisions to assess their impact and risk to ITSM activities if those decisions were to take place

✓ A means for modeling the breaking point at which the quality of ITSM practices becomes untenable.

In short, this tool may be used to support ITSM reporting and to model the impact of changes to the IT infrastructure or future business decisions.

Installing the Model

The model is built as a Microsoft EXCEL Spreadsheet and is included with the download files described in Chapter 1. Simply download the **Tooling Aid—ITSM Metrics Modeling Tool v3.xls** file to a desired folder on your PC or device. The PC itself should be running WINDOWS or other platform compatible with Microsoft Office.

It is recommended that you install the original version of the file and make changes only to copies of it. This will allow you to continually reuse the original to create baselines or future state models of your ITSM environment. For example:

1. Download and copy the original file to your PC or device as **Tooling Aid—ITSM Metrics Modeling Tool v3.xls**

2. Create a baseline of your ITSM current state environment by making a copy of this file, applying your metrics results and storing it as **ServiceBaseline.xls** (for example)

3. Create state models of your future ITSM and business decisions by creating copies of your baseline model and storing it with some relevant name (i.e. **PostMerger.xls** for example) and then apply changes to that baseline. In this way, you can create multiple versions of models based on different business scenarios and compare their risks and impacts. There will be more on this later.

When you first use the model, the values that are in the Success and Warning sections (colored grey) are arbitrary. These have been put in there for placeholders. These should be replaced with threshold levels that you wish to operate with.

How to Use the ITSM Metrics Modeling Tool

The model is simple to use. It consists of an EXCEL Workbook with individual Worksheets for many ITSM processes plus Service Desk and Workforce Worksheets. For each Worksheet, the steps are:

1) Fill out the **Success** and **Warning** sections with of each worksheet with the threshold values that wish to operate with for each Key Performance Indicator (KPI)

2) Fill out the **Data** (Operational Metrics) section of each worksheet with values from ITSM tools, reports and observations.

The model will automatically calculate the KPI values, compare them to the Success/Warning thresholds and derive a LOW, MEDIUM or HIGH score with corresponding color (Green, Yellow, or Red) to indicate target status. A Green color indicates the KPI is at a success target or better. A Yellow color indicates the KPI is between the success and warning targets. A Red color indicates the target is above or below the warning threshold that you set.

Critical Success Factors (CSFs) are automatically calculated based on the KPI values. A CSF consists of one or more KPIs that relate to it. It is color coded based on how well the combination of its KPIs were averaged. Therefore, a Red color indicates the CSF is at high risk, Yellow at Medium risk or Green at Low risk.

Outcome risk areas (Legal Exposure, Service Outages, Rework, etc.) are given a risk assessment (High, Medium or Low) based on the CSFs that apply to them. For these, the risk level colors indicate the possible likelihood that exists for each risk occurring.

CSFs are then factored into the process balanced scorecard dashboard section. The balanced scorecard data points (Customer, Capabilities, Operational, Financial and

Regulatory) are derived from one or more combinations of CSFs that impact them.

The data flows within the model are fairly simple. It can best be described using the metrics model previously presented:

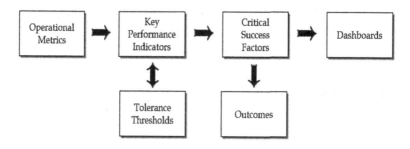

Figure 2: ITSM Metrics Model

1. Success and Warning thresholds are first entered for each process to describe acceptable and not acceptable KPI performance levels

2. Operational metrics are then entered for each process with live data from ITSM process reporting and other infrastructure measurements and observations

3. Key Performance Indicators (KPIs) are then calculated from the above and coded Green, Yellow or Red depending on how they fell within the specified success and warning thresholds

4. CSF risk levels are then calculated from combinations of KPI results and color coded as Green (Low), Yellow (Medium) or Red (High)

5. The process Balanced Scorecard Dashboard is then calculated from combinations of CSF results

Interpreting the Model Results

There are four items of interest that are output from this tool:

- ✓ KPI results
- ✓ CSF Results
- ✓ Balanced Scorecard Results
- ✓ Risk Assessment Results

KPI Results

These are the "Metrics That Matter".

The results for each KPI are shown as calculated from the Operational Metrics that were input compared to the success and warning threshold values. Color coding is based on how well the KPI fell within the Target and Warning threshold levels. Red results indicate potential areas that need to be improved.

As an example, the *Change Success Rate* KPI is calculated as follows from the Operational Metrics:

Number of Failed Changes /Total Changes Implemented

Therefore, if you implemented 1,000 changes and had no failures you scored a 100%. If you had 100 changes fail, you scored a 90%. This result is compared against what you input for target and warning threshold levels. If you indicated that your target was 98% and the warning level was 85% (as an example), then a 100% score would appear green. The 90% score would appear yellow. If your score ended up as, 84% for example, the score would appear red.

CSF Results

These are derived from specific KPIs that relate to them.

The CSF "Protect Services When Making Changes" is calculated from the following KPIs:

- ✓ Change Success Rate
- ✓ Emergency Change Rate
- ✓ Unauthorized Change Rate
- ✓ Change Incident Rate
- ✓ Change Management Process Maturity

These KPIs were chosen because they relate to specific threats to "Protecting Services When Making Changes". The model examines each of those KPIs and then provides a result equal to the KPI with the highest risk. Therefore, for example, this result could score a Red (High) in the situation where a low Emergency and Unauthorized Change rate existed with a high Change Incident Rate.

Risk Assessment Results

Risk assessment results are also included with each process. These represent outcomes derived from the KPIs and CSFs.

Each risk area represents a possible outcome based on how well CSFs were met. The color coded box to the right of each outcome indicates the likelihood that the associated risk might occur.

The ones and zeroes to the right identify which outcomes apply to each CSF. A starter set of values has been provided, but you may change these as you see fit. A one indicates that the outcome applies to the CSF (indicated in the above column header); a zero indicates that the outcome does not apply to the CSF.

Balanced Scorecard Results

A process Balanced Scorecard is presented that is simply a radar chart showing of each scorecard area (Customer, Capability, Operational, Financial and Regulatory). The data points represent an average of the risk levels for the risks that apply to each balanced scorecard area.

Modeling Business Decisions

One of the main purposes of the tool is to model the impact of business decisions or ITSM improvements that you are thinking about or planning to make. Examples of questions you may be trying to answer might be as follows:

✓ What will be the impact on our IT service quality if we put a major new application into production?

✓ How much operational risk will occur if we go through with a planned merger or acquisition?

✓ Which ITSM improvement initiatives will provide us with the most benefit?

✓ How many problems, incidents or changes can we handle before the quality of our services breaks down?

✓ What is the impact of increasing our Change Management CMMI Process Maturity from 2.4 to 3.5?

If using the tool to model things like this, you will first create a Baseline ITSM Model. The Baseline will represent your Current State practices. This model will be populated with the results of the way you currently utilize ITSM processes and activities. In this, you will populate the Tolerances and Operational Metrics as described earlier.

The next step is to create a series of Future State ITSM Models. These are also known as "What-If" Models. For each of these models you will make a replica of your baseline model and then make various changes to it that reflect various scenarios that you would like to model.

As an example, let's say that most of your ITSM Processes are at a 2.0 level. What might be the impact if you raise this level to 3.0? For this, you would:

1. Make a replica of your baseline ITSM model

2. Change the Process Maturity to 3.0 in the Operational Metrics section of each process you are interested in

3. View the results

You could then save this model for future reference and build other models to reflect other scenarios such as:

✓   What happens if the volumes of Incidents are decreased by 20%?

✓   What happens if the Emergency Change rate rises by 30%?

✓   What happens if Change Management labor is decreased by 10%?

Remember that only the Operational Metrics will be changed for What-If models. The model will calculate whether those changes resulted in KPIs that fell out of (or into) desired Tolerance threshold levels.

Chapter
23

Implementing an ITSM Metrics Program

ITSM Metrics Program End State

Implementing an ITSM Metrics Program is a key effort towards supporting the ITSM continuous improvement cycle. Remember, what can't be reported on can't be improved. A suggested vision for an ongoing ITSM Metrics Program will include actions that take place on a scheduled periodic basis to:

- ✓ Align metrics to current business need
- ✓ Report on CSF and KPI results
- ✓ Review and analyze those results
- ✓ Identify results that fall below performance goals
- ✓ Initiate actions to bring failing results back to acceptable performance levels.

It is highly recommended that these actions take place monthly if at all possible, but no less frequently than once every business quarter.

Key roles that will be needed to execute the metrics program include the following:

Metrics Program Leader

This role will be responsible for the overall ITSM Metrics Program. It ensures metrics are collected, produced and reported on within acceptable timeframes. It acts as a single point of contact to executive management for program results and actions being taken.

Process Owner

The Process Owner role is responsible for one or more ITSM processes. Its mission is to ensure the process is running efficiently, cost effectively and meeting business goals. In the context of the ITSM Metrics Program, this role is also responsible for providing the needed Operational and success/warning thresholds, identifying and initiating improvement actions if results fall below those threshold levels.

Metrics Analyst

This role is responsible for collecting, aggregating, summarizing and reporting on metrics data. It coordinates all activities needed to produce the metrics reports. It may also coordinate activities to produce one-time drill-downs or other views of data needed to support an improvement analysis effort. In addition, this role also maintains a repository of historical metrics information and reports.

Program Sponsor

This role covers the key senior executive who has approved and funded the overall metrics program.

Program Stakeholder

This role represents the customer of the metrics program. The stakeholder will review the metric results, identify concerns and issues, and agree on improvement actions when needed. Should results fall below Tolerance threshold levels, it is this role that will prioritize, agree and fund actions to make improvement efforts happen.

Tool Architect

This role maintains the overall program tooling solution architecture and ensures that tooling solutions integrate properly, provide accurate data, report and align with business need. In addition, this role may also be consulted for improvements and enhancements to tooling solutions that support the program over time.

Tool Technician

This role is responsible for maintaining and supporting the current tooling infrastructure that supports the metrics program. It may also be called upon to implement tooling changes to meet requirements for specific reporting needs.

Technical Writer

This role is responsible for documenting Program training materials, process, procedure and architecture guides.

A brief synopsis of day-to-day life in the metrics program at each reporting period might look like the following:

Align metrics to current business need

In this task, the team gets together with metrics stakeholders and identifies any needed changes to the metrics program. Are the current metrics still sufficient? Are changes needed to CSFs and KPIs? Is the current quality of reporting efficient? Changes are identified and scheduled to be in place for future reporting periods.

Report on CSF and KPI results

In this task, the team assembles aggregates and summarizes metrics data and assembles them into reports and/or dashboards. The reports are then distributed or placed online where they can be accessed and viewed. Metric information and reports are also stored for historical purposes in a metrics repository.

Review and analyze those results

In this task, the Program Leader, Process Owners, and the Metrics Analyst get together and take an initial pass at the results to determine that they are fair, accurate and representative of what took place during the previous reporting period.

Identify results that fall below performance goals

The Process Owners check to see which KPIs and CSFs have fallen below acceptable performance levels (success/ warning threshold levels). For each KPI, the Process Owners will develop action plans and alternatives to bring those results back to accepted levels. Approaches for doing this are then summarized for senior executive management.

Initiate actions to bring failing results back to acceptable performance levels

In this task, the Program Leader and Process Owners meet with metrics stakeholders and executive senior management to go over improvement action alternatives. Examples of outcomes of such a meeting might include:

- Agree and fund one or more projects to address needed improvements to correct the deficiencies found
- Adjust tolerance levels to reflect closer reality to what can actually be delivered and achieved
- Agree to monitor deficiencies for a set period of time and delay actions to make sure deficiencies are not one-time events.

A summary of roles and responsibilities for the ITSM Metrics Program as suggested is presented below *(Note: the Technical Writer is not typically needed for operating the program, only building it)*:

Table 67: Metrics Program Roles and Responsibilities

Key Task	Program Leader	Process Owner	Metrics Analyst	Program Sponsor	Stakeholder(s)	Tool Architect	Tool Technician
Align metrics to current business need	A	R	C	S	C	C	I
Report on CSF and KPI results	A	C	R	S	I	I	C
Review and analyze those results	R	A	C	S	I	I	I
Identify results that fall below performance goals	A	R	C	S	I	I	I
Initiate actions to bring failing results back to acceptable performance levels	R	A	I	C	S	C	C

S = Signatory, A = Accountable, R = Responsible, C = Consults, I = Informed

The importance of having an ongoing ITSM Metrics Program cannot be understated. Without this, it is impossible to accurately identify needed service improvements on a timely basis.

Without this, IT cannot demonstrate that it can effectively govern itself and align its priorities with what is needed for the business.

Work Breakdown Structure

The suggested Work Breakdown Structure that represents the key deliverables of an ITSM Metrics Program Implementation Effort can be shown as follows:

Figure 3: ITSM Program Work Break Down Structure

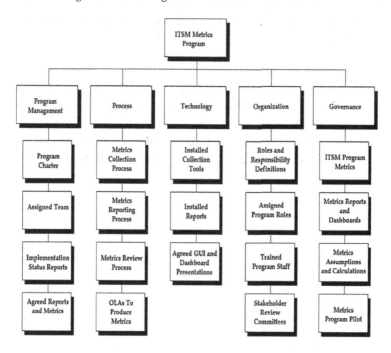

The above deliverables are divided into 5 project tracks that can work in parallel:

Program Management
Oversees and manages the implementation effort

Process
Develops processes needed to operate and run the program on an ongoing basis. This also includes establishing OLA (Operational Level Agreements) as needed to obtain metrics information or produce reports.

Technology

This work track is responsible for developing the program solution tooling architecture. It also is responsible for procuring, installing and testing of those tools.

Organization

This work track is responsible for identifying and documenting program roles and responsibilities. It also ensures that roles have been assigned to program staff once they are approved.

Governance

This work track is responsible for identifying the needed metrics, documenting their assumptions and calculations. It also oversees testing of the solution and will establish and manage a pilot effort. Governance will have the key voice in determining when the metrics program is ready to go live.

Implementation Approach

The suggested work approach uses 5 key work steps to implement the ITSM Metrics Program:

1. Program Initiation
2. Program Design
3. Program Building and Testing
4. Program Pilot
5. Program Rollout and Transition

Ideally, these work steps should be incorporated into an overall ITSM Process Implementation Program as part of the Governance tasks for that effort. For the purposes of this book, we will assume that this program is being implemented standalone to better outline suggested tasks that need to be completed.

It is suggested that the Program be steered by and follow the ITSM Release and Change Management processes. A more detailed explanation of each step is as follows:

Program Initiation

This set of tasks involves developing the Program Charter and preliminary scope for the effort. The preliminary scope will outline what business units are covered by the effort, key CSFs and KPIs that are initially desired, who will receive and act on metrics reporting and any key assumptions. This effort ends with an approved program implementation project, assigned project lead and appropriate funding needed to complete its goals.

A list of tasks for this work step is as follows:

Table 68: Program Management Roles and Responsibilities

Initiation Work Tasks	Program Leader	Process Owner	Metrics Analyst	Program Sponsor	Stakeholder(s)	Tool Architect	Tool Technician	Technical Writer
Program Management								
Develop Preliminary Program Scope	A	C	I	C	C	I	I	I
Build ITSM Metrics Program Charter	A	C	I	C	C	I	I	I
Approve ITSM Metrics Program Charter	A	C	I	S	C	I	I	I
Assign Implementation Team	A	C	I	S	C	I	I	I
Build Program Work Plan	A	C	C	S	C	C	C	C
Conduct Program Kickoff Meeting	A	C	C	I	C	C	C	C

(S=Signatory, A=Accountable, R=Responsible, C=Consults, I=Informed)

Program Design

In this work step, design tasks take place to identify the metrics that will be used, definitions, key assumptions and calculations. Identification of how each metrics will be sourced, what reporting tools, report and dashboard templates will also take place. Processes for collection, and reporting will be designed and roles and responsibilities for these will also be identified.

A list of tasks for this work step is as follows:

Table 69: Program Design Roles and Responsibilities

Design Tasks	Program Leader	Process Owner	Metrics Analyst	Program Sponsor	Stakeholder(s)	Tool Architect	Tool Technician	Technical Writer
Governance								
Identify ITSM Program Metrics	A	C	R	S	C	C	I	I
Define Assumptions and Calculations	A	C	R	C	C	C	I	I
Design Metrics Reports and Dashboards	A	C	C	S	C	C	C	I
Agreed GUI and Dashboards	A	C	C	S	C	C	C	I
Organization								
Define Program Roles and Responsibilities	A	C	C	S	C	C	I	I
Define Program Skills	A	C	C	S	C	C	I	I
Identify Program Training Requirements	A	C	C	S	C	C	C	I
Identify Stakeholder Review Committee	A	C	I	S	C	I	I	I
Develop Program Communications Plan	A	C	I	S	C	I	I	I

Design Tasks	Program Leader	Process Owner	Metrics Analyst	Program Sponsor	Stakeholder(s)	Tool Architect	Tool Technician	Technical Writer
Process								
Develop Metrics Collection Processes	A	C	R	I	C	C	C	I
Develop Metrics Reporting Process	A	C	R	I	C	C	C	I
Develop Review Committee Agenda	A	C	R	C	C	I	I	I
Develop Metrics Review Process	A	C	R	I	C	I	I	I
Draft Metrics Production Guide	A	C	C	S	C	C	C	R
Technology								
Identify Metric Sourcing Strategy	A	C	C	I	C	R	C	I
Identify Metrics Repository Architecture	A	C	C	I	C	R	C	I
Identify Metrics Collection Tools	A	C	C	I	C	R	C	I
Identify Metrics Reporting Tools	A	C	C	I	C	R	C	I
Identify Metrics Presentation Tools	A	C	C	I	C	R	C	I
Drafty Metrics Architecture	A	C	C	S	C	R	C	C
Agreed GUI and Dashboards	A	C	C	S	C	R	C	I
Program Management								
Update Implementation Plans	A	C	C	C	C	C	C	C
Integrate Project Activities	A	I	I	C	C	I	I	I
Report Program Status	A	C	C	C	I	C	C	C

Program Building and Testing

In this work step, tasks are done to build and test the Program tools and architecture. This includes procuring and assembling needed tools and documenting detailed procedures for metrics collection and reporting. It also includes unit and integration testing of program tools and processes.

It is in this step that a Pilot be selected for the Program. The Pilot may consist of:

- ✓ One or more business units that will be covered under the metrics program for a set period of time
- ✓ Covering all business units in a "trial" mode for a set period of time
- ✓ Some combination of the above

The purpose of the Pilot is to ensure that the Program is operating as planned and that acceptance for it is gained. Therefore, expectations with Pilot candidates should be clearly set to let staff know that some errors and issues may occur during this time. Without this, a perception may develop that the Program is flawed. These kinds of communications need to be established in advance as part of this work effort.

A list of tasks for this work step is as follows:

Table 70: Build and Test Roles and Responsibilities

Build and Test Tasks	Program Leader	Process Owner	Metrics Analyst	Program Sponsor	Stakeholder(s)	Tool Architect	Tool Technician	Technical Writer
Governance								
Develop Test Conditions and Results	A	C	R	C	C	C	C	I
Oversee and Review Test Results	A	C	R	C	C	C	C	I
Select and Agree Pilot	A	C	R	S	C	C	C	I
Draft Pilot Plan	A	C	C	C	C	C	C	I
Oversee Training For Pilot Staff	A	C	C	I	I	I	I	I
Organization								
Update Job Descriptions For Roles	A	C	C	S	C	C	C	I
Create Program Training Materials	A	C	C	S	C	C	C	R
Assign Personnel For Pilot	A	C	C	S	C	C	C	I
Conduct Program Training For Pilot	A	C	C	I	I	C	C	C
Process								
Develop Detailed Procedures	A	C	R	I	C	C	C	I
Test Processes and Procedures	A	C	R	I	C	C	C	I
Update Metrics Production Guide	A	C	C	C	C	C	C	R
Technology								
Procure Program Tools	A	C	C	S	C	R	C	I
Install and Customize Tools	A	C	C	I	C	R	C	I
Test Program Architecture	A	C	C	I	C	R	C	I

Build and Test Tasks	Program Leader	Process Owner	Metrics Analyst	Program Sponsor	Stakeholder(s)	Tool Architect	Tool Technician	Technical Writer
Validate Program Reports/ Presentations	A	C	C	S	C	R	C	I
Program Management								
Update Implementation Plans	A	C	C	C	C	C	C	C
Integrate Project Activities	A	I	I	C	C	I	I	I
Report Program Status	A	C	C	C	I	C	C	C

Program Pilot

In this work step, the Program Pilot is executed. This involves running and operating the Pilot as if the Program were running live in real production. The Pilot should be monitored for deficiencies. It is important that these are addressed right away when discovered or the effort will appear flawed. During this period, the Program team is executing two main tasks: supporting the Pilot and making preparations for final rollout of the Program solution.

A list of tasks for this work step is as follows:

Table 71: Program Pilot Roles and Responsibilities

Pilot Tasks	Program Leader	Process Owner	Metrics Analyst	Program Sponsor	Stakeholder(s)	Tool Architect	Tool Technician	Technical Writer
Governance								
Oversee and Manage Pilot Results	A	C	C	C	C	C	C	I
Maintain Inventory of Deficiencies	A	C	R	C	C	C	C	C
Ensure Deficiencies Addressed	A	C	C	C	C	C	C	C
Draft Program Rollout Plans	A	C	C	C	C	C	C	I
Agree Program Rollout Plans	A	C	C	S	C	C	C	I
Organization								
Monitor Pilot For Skill Levels	A	C	C	I	C	C	C	I
Address Training Changes If Needed	A	C	C	C	C	C	C	C
Manage Pilot Stakeholders	A	C	I	C	C	C	I	I
Process								
Monitor Pilot For Process Deficiencies	A	C	C	I	C	I	I	I

Pilot Tasks	Program Leader	Process Owner	Metrics Analyst	Program Sponsor	Stakeholder(s)	Tool Architect	Tool Technician	Technical Writer
Address Process Deficiencies If Found	A	C	R	C	C	C	C	I
Support Pilot Processes	A	C	C	C	C	C	C	I
Technology								
Monitor Pilot for Technical Deficiencies	A	C	C	I	C	R	C	I
Address Technical Issues If Found	A	I	I	I	I	R	C	I
Support Pilot Technical Architecture	A	I	I	I	I	R	C	I
Program Management								
Update Implementation Plans	A	C	C	C	C	C	C	C
Integrate Project Activities	A	I	I	C	C	I	I	I
Report Program Status	A	C	C	C	I	C	C	C

Program Rollout and Transition

In this work step, the ITSM Metrics Program is rolled out to the rest of the organization. This involves executing the rollout plans. Operational tasks for the program are converted to production and remaining additional staff (if needed) are trained and put into place.

At the end of this step, the ITSM Metrics Program should be fully operational. The only additional task at this point is to initiate ongoing review and monitoring of the Program to identify future improvement enhancements. The suggested approach described for implementation can be reused in abbreviated form as needed to implement improvements over time.

A list of tasks for this work step is as follows:

Table 72: Rollout and Transition Roles and Responsibilities

Rollout and Transition Tasks	Program Leader	Process Owner	Metrics Analyst	Program Sponsor	Stakeholder(s)	Tool Architect	Tool Technician	Technical Writer
Governance								
Oversee and Manage Program Rollout	A	C	C	S	C	C	C	I
Maintain Inventory of Deficiencies	A	C	R	C	C	C	C	C
Ensure Deficiencies Addressed	A	C	C	C	C	C	C	C
Validate Program Goals Achieved	A	C	C	S	C	C	C	I
Identify Future Improvements	A	C	C	C	C	C	C	I
Organization								
Conduct Planned Training Activities	A	C	C	I	C	C	C	I
Identify Training Improvements	A	C	C	C	C	C	C	I
Manage Program Stakeholders	A	C	C	C	C	C	C	C
Process								
Monitor Program For Deficiencies	A	C	C	I	C	I	I	I
Identify Process Improvements Needed	A	C	R	C	C	C	C	I
Technology								
Support Program Technical Architecture	A	I	I	I	I	R	C	I
Identify Future Technical Improvements	A	I	I	I	I	R	C	I

Rollout and Transition Tasks	Program Leader	Process Owner	Metrics Analyst	Program Sponsor	Stakeholder(s)	Tool Architect	Tool Technician	Technical Writer
Program Management								
Report Program Status	A	C	C	C	I	C	C	C
Close Project	A	I	I	S	I	I	I	I

About the Author

 **Randy A. Steinberg** has extensive IT Service Management and operations experience gained from many clients around the world. He authored the ITIL 2011 Service Operation book published worldwide. Passionate about game changing management practices within the IT industry, Randy is a hands-on IT Service Management expert helping IT organizations transform their IT infrastructure management strategies and operational practices to meet today's IT challenges.

Randy has served in IT leadership roles across many large government, health, financial, manufacturing and consulting firms including a role as Global Head of IT Service Management for a worldwide media company with 176 operating centers around the globe. He implemented solutions for one company that went on to win a Malcolm Baldrige award for their IT service quality. He continually shares his expertise across the global IT community frequently speaking and consulting with many IT technology and business organizations to improve their service delivery and operations management practices.

Randy can be reached at RandyASteinberg@gmail.com.